Anonymous

Arctic Geography and Ethnology

A selection of papers on Arctic geography and ethnology

Anonymous

Arctic Geography and Ethnology
A selection of papers on Arctic geography and ethnology

ISBN/EAN: 9783337323981

Printed in Europe, USA, Canada, Australia, Japan

Cover: Foto ©berggeist007 / pixelio.de

More available books at **www.hansebooks.com**

Arctic Geography and Ethnology.

A SELECTION OF PAPERS

ON

ARCTIC GEOGRAPHY AND ETHNOLOGY.

REPRINTED, AND PRESENTED TO

THE ARCTIC EXPEDITION OF 1875,

BY

THE PRESIDENT, COUNCIL, AND FELLOWS OF THE
ROYAL GEOGRAPHICAL SOCIETY.

LONDON:
JOHN MURRAY, ALBEMARLE STREET.
1875.

PREFACE.

THE President and Council of the Royal Geographical Society suggested that a selection of papers on various branches of science relating to the Arctic Regions, which are rendered inaccessible through being bound up in 'Transactions' and 'Proceedings' with other irrelevant matter, should be reprinted for the use of the Arctic Expedition. This suggestion, so far as regards subjects other than geography and ethnology, was adopted by the Admiralty on the recommendation of the Council of the Royal Society, and a collection of papers and extracts from books on zoology, geology and physics, will be reprinted at the public expense for the use of the expedition.

The present volume contains a series of papers on Arctic geographical and ethnological subjects, which it was thought might be useful to the officers of the expedition; and which has been prepared by a Committee appointed by the Council, and at the expense of the Royal Geographical Society. It is a contribution presented to the Arctic Expedition by the Society, in the hope that some use and instruction may be derived from it, and with the warmest and most heartfelt wishes for the success and safe return of the explorers, on the part of the Council and Fellows.

The Volume is divided into two sections—on Geography and Ethnology.

The first series of papers in the Geographical Section is by Dr. Robert Brown, F.R.G.S., who has twice visited Greenland, and who is one of the highest living authorities on all scientific subjects connected with that region. Dr. Brown, after briefly describing the Greenland coast-line, gives an account of all the different attempts that have been made to penetrate into the interior. He then treats of the Greenland glacier system, of the action of sea-ice, of the rise and fall of the coast, and of the formation of fjords, and concludes with some speculations on the northern termination of Greenland, and on debateable points regarding the physical structure of the vast icy continent.

Dr. Brown's series is followed by three papers reprinted from the 'Journal' of the Royal Geographical Society. The first, by Baron von Wrangell, is interesting, as being the first proposal to attempt to reach the Pole by the route of Smith Sound. The second is a valuable criticism on the narrative of Dr. Kane's discoveries, by Dr. Rink, the eminent Danish Naturalist, and Director of the Greenland Board of Trade; and the third is a paper on the Arctic Current around Greenland, by the Danish Admiral Irminger.

The concluding series of Papers, in the Geographical Section, is by Admiral Collinson. The full results of that distinguished officer's remarkable Arctic voyage have never been given to the public; and both the Fellows of the Society and the officers of the Arctic Expedition are to be congratulated in having, on this occasion, elicited so valuable an instalment. Admiral Collinson gives his notes on the state of the ice, and on indications of open water, from the mouth of the Siberian river Kolyma, along the shores of Arctic America, to Bellot Strait. He also furnishes a narrative of all the expeditions that have explored the shores of Arctic America from Point Barrow to the Mackenzie River, and from the Mackenzie to the Back River, including his own voyage, and concludes with some general observations on the ice.

The Ethnological Section commences with two papers on the origin and migrations of the Greenland Eskimo, and on the Arctic Highlanders. Then follows a sketch of the Eskimo grammar, and a series of classified vocabularies taken from the lists of Egede, Kleinschmidt, Janssen, and Admiral Washington.

The compilation of the list of names of places in Greenland has been a difficult task, and it is feared that it falls short of what might have been prepared if more time could have been bestowed upon it. The intention has been to give the name of every place on the coast of Greenland from the Dannebrog Islands, in latitude 65° 15′ N. on the eastern side, round Cape Farewell, to the entrance of Smith Sound; with columns for the Eskimo names, their meanings, identifications of ancient Norman sites, Danish names, names and latitudes on the Admiralty Chart, and remarks. The Eskimo meanings have been kindly supplied by Dr. Rink, and the Norman identifications are mainly due to the learning of Mr. Major. Much laborious assistance, in the preparation of this list, has also been given by Commander A. H. Markham, R.N., F.R.G.S., of H.M.S. *Alert.*

A short but interesting paper follows, by Dr. Rink, on the descent of the Eskimo; and the elaborate memoir by the late Dr. Simpson, R.N., of H.M.S. *Plover*, on the Western Eskimo, completes this Section. The volume concludes with a report, and a series of questions, which were prepared in 1872, by a Committee of the Council of the Anthropological Institute.

CLEMENTS R. MARKHAM,

SECRETARY R.G.S.

TABLE OF CONTENTS.

ETHNOLOGY.

I. PAPERS ON THE GREENLAND ESKIMOS. BY CLEMENTS R. MARKHAM.

PAGE

II. ON THE DESCENT OF THE ESKIMO. BY DR. RINK .. 230

III. THE WESTERN ESKIMO. BY DR. SIMPSON, R.N. .. 233

IV. REPORT OF THE ANTHROPOLOGICAL INSTITUTE.. .. 276

Questions for Arctic Explorers 281
  1. General. By Dr. Barnard Davis 281
  2. Religion, Sociology, &c. By Mr. Tylor .. 282
  3. Remains of Ancient Tribes. By Mr. Boyd
    Dawkins 283
  4. Customs relating to War. By Colonel Lane
    Fox 283
    Arrow-Marks and other Signs 286
    Drawings, Ornamentation, &c. 287
  5. Further Ethnological Questions. By Mr.
    Franks 288
  6. Physical Characteristics. By Dr. Beddoe .. 290
  7. Further Ethnological Inquiries. By Dr.
    Turner 291
  8. Suggested Inquiries. By Capt. Bedford Pim 292

MAPS.

Map drawn by Erasmus York, the "Arctic Highlander" .. *to face* 184

Map of the part of the Coast of Greenland containing the
ancient Norman Settlements .. .. ,, 209

GEOGRAPHY.

I.

ON THE PHYSICAL STRUCTURE OF GREENLAND.

In drawing up a summary, as brief as such a wide subject will admit of, regarding our knowledge of the physical structure of Greenland, apart from its geology, which will be found in the contributions to the Manual prepared by the Arctic Committee of the Royal Society, the materials at my disposal will be best utilized by adopting the following division :—(1) A description of the coast. (2) A summary of what we know of the interior and of the chief attempts which have been made to penetrate it. (3) The Greenland ice and Greenland glaciers. (4) The nature of the Greenland fjords. (5) A discussion of the question regarding the probable termination of Greenland; and, finally, (6) a few memoranda may be added in regard to the points discussed in the preceding pages, in reference to which our knowledge is still imperfect, but which the researches of the present Expedition could do much to solve.

1. The Greenland Coast-line.

The Admiralty chart, and the numerous elaborate ones in the narrative of the German Expedition to East Greenland,[1] are so detailed, that any minute description of the coast-line is superfluous. I will, therefore, merely confine myself to a brief outline, more as connecting the topographical portion of my subject with that which is more purely physico-geographical, than with any view to supply what a glance at the chart will much more efficiently afford the reader.

[1] 'Die Zweite Deutsche Nordpolarfahrt' (Leipzig, 1872-75), or its partial translation (but without all the maps, &c.) by Messrs. Bates and Mercier (London, 1874).

B

The general form of Greenland, as at present portrayed on our maps, is roughly triangular. It is probable that further discoveries on the northern shores will show it to be more ellipsoidal than triangular in shape. Its interior is unknown, but its shores on the western and part of the eastern sides have been more or less completely explored. At almost no place is there a straight or unbroken line of coast; deep fjords, at short intervals, running more or less parallel with each other, often for great distances into the land, and in some cases divided into numerous branches or tributary fjords, intersect the coast. These fjords are much more numerous on the west than on the east coast—a fact which we shall see is true of every other region where these peculiar intersections of a coast-line are found (p. 69); and this fact may be received as some ground for the belief that the inland ice (p. 30) which covers the whole interior of the country slopes more to the western than to the eastern sides.

1. *The East Coast.*—The Spitzbergen Ice-stream—a broad river of pack-ice, floes, &c.—is carried by the current from the direction of Spitzbergen down the eastern shores of Greenland, south at least of lat. 64°, and is drawn up Davis Strait by the in-draught of the water until it impinges on the coast about the vicinity of Holsteensborg. In this current are brought great quantities of drift-wood, which has passed out of the mouths of the Siberian rivers, and white bears, which afford a lucrative object of chase to the South Greenlanders.

The most northerly point on the coast of Greenland which has ever been sighted, is the mythical land which is said to have been visited by Lambert in 1670. But this record is so dubious, that we may really set down the furthest northern point reached by the German Expedition on the 15th of April, 1870, viz., Cape Bismarck, or a little beyond, in lat. 77°, as the limit of our knowledge of the eastern shores. South of that parallel the coast-line has been partially laid down by Scoresby, and by the expedition mentioned, until we come to lat. 69° 12', near Knighton Bay, when again the chart fails us. Between the points mentioned the coast is broken by fjords and bays, with numerous off-lying islands. The most extensive of these fjords is that of the Kaiser Franz Joseph, a beautiful inlet (with many tributaries), which stretches into the interior for an unknown distance.

Scarcely less beautiful are Ardencaple Inlet and the Fligely and Tyrolese Fjords, though neither is equal in extent or grandeur to that named in honour of the Austrian Emperor. Koldewey's, Clavering's and Shannon islands form the greatest extent of detached land.

Petermann's Peak (14,000 feet), and Payer's Peak (7600 feet), are the highest points of land in that region. In Greenland, it may be remarked, there are few high elevations. "Greenland's icy mountains" are to some extent a hymnal myth! Scoresby's Sound is an unexplored inlet of perhaps an extent even greater than any of those named. Davy Sound may also prove to be an extensive northern tributary of Scoresby's Sound.

South of Knighton Bay, until we come to the White Saddle Island in lat. 65°, we may be said to know nothing of the coast. Here and there a cape has been sighted and a name applied to it; and practically a dotted line might fitly express all the exact knowledge which we possess in regard to it. From lat. 65° to Cape Farewell, the southern termination of the country, the coast has been laid down from the sketches of the old Norsemen, and from the observations of Graah and others, who went in search of the "lost colonies," believed, but erroneously, to have been situated, up to the period of the Middle Ages, on the south-eastern portion of Greenland.[1] The coast-line is broken by fjords, with very few islands lying off their mouths.

2. *West Coast.*—Cape Farewell (called by the Greenlanders *Kangekyadlek*, or the cape running to the westward) is on a small island (Sermilik). From this point up lat. 73° 40′ (Tessiussak or Kingatok) the coast has been more or less perfectly surveyed. Of the southern harbours and inlets we indeed possess some excellent charts by the Danish naval surveyors. At all events, no important points in its geography are unknown, and it may be said that, for all geographical purposes, the west coast of Greenland is perfectly well known within the limits of the Danish possessions. Its general character is much the same as the rest of the Greenland continent —not overlaid by ice—and will be described, so far as the nature of this memoir requires, in a subsequent section (p. 29). Sukkertoppen ("the sugar-loaf") and Sanderson's Hope (Kasorsoak) are about the highest points of the coast.

North of these limits the unexplored or imperfectly known region commences. The bottom of Melville Bay is, for instance, entirely unknown. Great glaciers, fjords, and islands—one of which is said to constitute that pillar-like land to the entrance of the bay known as the Devil's Thumb—will most likely be found to be the prevailing character of the coast. The bottom of few, if any, of the inlets north of this are known, and the outer coast-line very imperfectly. How much, or how little, we know of Smith

---

[1] The " Öster Bygd " has now been proved to have been on the west coast.

Sound the charts and other documents will have so fully explained to the Expedition, that it is manifestly out of the province of the present writer to enter upon this subject. The physical characteristics of the country do not, so far as a mere study of the published sources of information which we possess in regard to it will allow us to judge, differ in any remarkable degree from the region already spoken of.

2. The Interior of Greenland.[1]

The interior of any considerable tract of land has always a mysterious interest surrounding it, especially when its coasts have long been a familiar object on our maps. Indeed, now-a-days, when the broad features of the world, excepting those of some of the more remote Arctic and Antarctic regions, are tolerably well known, little remains to the geographical explorer but the investigation of the interior of some of the older continental masses of land. Even with his ambition so bounded, the traveller need not, like a second Alexander, sit down and weep because there are no more worlds to conquer. The geography and resources of scarcely any great mass of land, from Australia to Greenland, with the exception of the long civilised and inhabited European countries, are well known, and some even very near to the great centres of population and enterprise of the world, such as Iceland and Greenland, are little, if at all known, or even attempted to be explored. Yet the superficial area of Greenland cannot be less than 750,000 square miles—in a word, it is a continent.

It is now upwards of 1000 years[2] since the banished Iceland Viking, Red Erik (Thorwards' son), discovered the land to which he applied the somewhat *couleur-de-rose* name of " Grönland." For upwards of 700 years it was settled on its southern shores, or visited for hunting, fishing, or trading purposes, by his countrymen from Iceland and Norway. Thirteen bishops were ordained to preside over this frozen diocese, and churches and villages yet remain, in the shape of massive rude ruins, to attest how strong a hold it had taken on the colonising spirit of Scandinavia. For nearly 300 years exploring vessels of almost every European maritime nation have passed along its ice-bound shores, either for the purpose of exploring its northern termination or of tracing the trend of its unknown

---

[1] Condensed and re-cast, with corrections, from ' Das Innere von Grönland ' (Petermann's ' Geog. Mittheilungen,' 1871).
[2] This date is not certain; some authors give it as A.D. 983. See also Konrad Maurer's ' Island, von seiner ersten Entdeckung bis zum Untergange des Freistaats ' (1874).

eastern coast. For upwards of 200 years thousands of English,
Dutch, Danish, German, Norwegian, American, or French ships
have visited it, hunted the whale and the seal in its waters, or every
summer battled with their giant quarry in the more distant seas
which wash its shores; and finally, it is now nearly a century
and a half ago since the Danish Government first established
trading-posts on the western coasts from near Cape Farewell to
almost 74° N. latitude, where reside from year to year educated
and intelligent Danish officers with the whole resources of the
trading monopoly at their disposal. Yet, as far as any definite
knowledge of the interior goes, we know almost as little to-day as
we did when Erikr Rauthri returned home again to Snæfjeld-
jokelsfjord, boasting of the new country he had discovered.[1]

True, we know that it is covered with an immense glacier expansion. But whether this glacier expansion is unbroken from north
to south or from east to west we can only reason from analogy, and
are not able to speak with the authority and confidence which actual
observation gives. Before we hastily vent our indignation in the
stereotyped phrase of "it being discreditable to the enterprise of
the age" that this should be, let us glance for a moment at the
causes of this. Though so near Europe, Greenland is yet in reality
far off, communication with it being rare and slow, while once
there, there is little to attract the attention of an explorer, who is
apt to think his time more profitably and pleasantly spent in more
fruitful and hospitable regions. Accordingly, while the mysteries
of Africa are explored at every risk of life and health, and the
eucalyptus-thickets of Australia never lack Englishmen and Germans willing to risk a grave among them, and the gorgeous wonders
of Amazonian vegetation attract men to wander in awe-struck
admiration amongst it, the icy interior of familiar Greenland lies
solitary, mysterious, and unknown. The Danish residents in
Greenland are too occupied with their duties, and, unless under
special encouragement from the Government, can scarcely be expected to undertake what has found no attractions for professional
geographers and explorers. When I said that it is *known* that the
interior is an ice-waste covered with a huge *mer de glace*, I ought
to have qualified this statement by saying that this is only a matter

---

[1] "It was a green land, a fair country, greener than Iceland," loudly in ale-
house and market-place proclaimed this lusty, boisterous, roystering drinker of
ol and mead. The fact is, that, in his own small way, this same banished son of
the banished son of Jadar, the Norwegian jarl, was a "promoter' of a joint-stock
company for colonization, and knew as well as anybody within the city of London
or elsewhere what was in a name. "For," quoth he, "if the land have a good
name, it will cause many to come thither."

of knowledge to those who have devoted attention to the subject,
for, to the ordinary geographer and naturalist, the fact does not
seem to be generally known.   It will, therefore, be useful to give
a summary of the different attempts—futile though most of them
have been—to penetrate the interior of the frozen land, and to
shortly sum up what the present state of our knowledge would lead
us to deduce regarding the structure and configuration of this interesting Arctic Continent.

1. *Ocean and Landorff's Attempt in* 1728.—As far as I can learn,
this is the first attempt made to penetrate the interior of Greenland,
and from the ignorance it displayed of the nature and character of
the country to be passed over, we may well suppose that it was
planned in a time of supreme unacquaintance with the existence of
the inland ice. Major Ocean[1] and Capt. Landorff were respectively
the governor and commandant designate of a fort which the Danish
Government proposed to establish on the east coast of Greenland.
They took with them an armed company, artillery and horses, from
Denmark. The horses died on the passage out; and so a grandly
planned expedition failed, owing to its having been projected in
utter ignorance of the nature of the country. Finding that it was
all but impossible, on account of the great ice-stream which is ever
pouring down that coast, to reach the seat of Government, these
gallant officers proposed what appears to us now almost too ludicrous and madcap a scheme to be seriously related : viz., to ride on
horseback across the country from the west to the east coast. We
must, however, remember that a century and a half ago little or
nothing was known about Greenland except by vague tradition or
the tales of the Eskimo, repeated by Hans Egede, who had just
established his trading mission eight years, and was but imperfectly
acquainted with the language of the Eskimo, and more than sus-
picious of their veracity. It is also as well to bear in mind that
some of the South Greenland fjords support a few cattle and sheep,
and, therefore, in some respects, justify the name which Erikr
Rauthri applied to the country when he first discovered it. They
seem to have attempted it on foot, some will even say on horseback; but history has preserved us but scanty details of this
extraordinary attempt, for all that I can find regarding it is a
doleful lament that the route taken was covered with glaciers and
chasms. Egede seemed to have been well acquainted with the
nature of the inland ice, for, in all the attempts either made by him

---

[1] According to my notes of the expedition. Nordenskjöld, however, in his
'Redogörelse för en Expedition till Grönland, år 1870,' gives the name as
" Paars."

or under his direction, we never found him attempting to cross the
country, but always to work laboriously round Cape Farewell.
Soon after this the expanse of the inland ice over the interior seems
to have been well known, for Cranz gives us a lucid description of
it: and Otho Fabricius, the celebrated naturalist and philologist,
who was in the country about the same period, describes, in his
'Fauna Groenlandica,' published in Copenhagen in 1780, the
interior in these words (page 4): "Interioribus ob plagam gla-
cialem continuam inhabitabilibus."

2. *Dalager's Journey in* 1751.—The Danish settlement of Fredrikshaab, situated in lat. 60° N., and long. 50° W. of Greenwich, was
founded in 1742 by a Danish merchant, Jakob Severin [1]—the trade
of Greenland not being then, as it is now, a strict monopoly of the
Government. The first traders were Gelmeyden and Lars Dalager,
men of much energy and rather celebrated in the simple annals of
Greenland. Lars appears to have been the author of a work on
Greenland,[2] which I have not seen, though there are quotations
from it, and from his private letters, both in the works of Cranz and
Saabye on Greenland. From the former of these we derive our
information regarding this enterprising attempt to penetrate to the
interior of the country. As it was one of the first, it probably yet
stands alone as one of the most interesting and energetic of all the
attempts which succeeded it. He informs us that on the 28th[3] of
August, 1751, he sent the great boat to search for firewood, north
of the "Iceblink,"[4] and a day's journey north of Fredrikshaab,
while he followed in his hunting-boat. A Greenlander had, in the
preceding month, pursued his game so high in the country that he
could see, as he said, the mountains of the ancient "Kablunaks," or
Europeans, who had in the middle ages settled in South Greenland.
Induced by this intelligence, he determined to seize the present
opportunity of attempting a passage to the east side. On the 2nd
of September, accompanied by the Greenlander, the Greenlander's
daughter, and three other natives, he set out on his tour from a
bay on the south of the "Iceblink." They tied their bag of
provisions and their furs to sleep in together, and gave them to the
girl to carry. The rest of the party took each his little skin kajak or
Greenland boat on his head, and a musket on his shoulder, and in

[1] Severin was the founder of several other settlements. His name is perpe-
tuated in "Jakobshavn," a settlement on the southern shores of Disco Bay.
[2] 'Grönlandske relationer, indeholdende Grönlandernes liv og levnet, deres
skikke og vedtägter, samt temperament og superstitioner; tillige nogle korte
reflexioner over missionen, sammenskrivet ved Fredrikshaab's Colonie i Grönland af *Lars Dalager*, Kjöbmand.'
[3] "Old style," I presume. [4] A projecting glacier in lat. 62° 30′ N.

this manner took up their march. The first half-mile was along
a brook-side, and was level and easy walking; but they had now a
high and rugged rock to cross, and frequently fell down with
their boats on their heads. By sunset they had reached a large bay
on the other side, fourteen leagues in length, a hard day's pull for
an expert rower. In former times the Greenlanders could row into
this directly from the sea, but, owing to many of the fjords having
become filled up by glacier-mud and ice, this cannot be done now.
The next day they launched their kajaks, and rowed for 4 miles
straight across the bay to the north side. They then left their boats
covered with stones and pursued their journey on foot to the north-
east. Crossing a ridge of rocks, they came in the evening to firm
ice. Early on the morning of the 4th, they set out over it to the
nearest mountains of the Iceblink, at about 4 miles distant. "The
road was as level as the streets of Copenhagen." An hour after
sunset, they arrived at the top. The next day they occupied in
hunting reindeer, one of which they killed, and the raw flesh of
which fell to the Greenlanders; for, as there was neither grass nor
brush to kindle a fire, Dalager was obliged to be satisfied with a
piece of bread and cheese. On the 5th they travelled about 4 miles
to the highest rock on the borders of the Iceblink, but were seven
hours on the road, as the ice was uneven and full of *crevasses*, which
obliged them to make frequent detours. About 11 o'clock they
came to the rock, and, after taking an hour's rest, began to ascend.
Towards 4 o'clock they gained the summit, spent with fatigue.
Hitherto they had only been travelling over the ground bordering
the great interior *mer de glace*, or over some defluent glaciers; but
now an extensive prospect burst upon their view on all sides,
striking them with wonder, particularly when the vast fields of ice
were seen stretching across the country in the east coast, bounded
in the distance by mountains whose tops were covered with snow
like those on which they stood. At first these mountains seemed
only 6 or 7 leagues distant, but when they looked towards Godthaab
(lat. 64° 10' 36" N., long. 51° 45' 5" W.) and saw the mountains in
its vicinity appear equally large though at least 100 miles off, they
were obliged to enlarge their estimate. The adventurers remained
till evening on the mountain-side, then descending a short way they
lay down to rest; but Dalager tells us that the activity of his
thoughts, aided by the cold, drove away sleep. On the morning of
the 6th they shot another reindeer close to their resting-place. All
scruples had now vanished, and, craving for something warm,
Dalager took a draught of its warm blood, which refreshed him
much, and joined the Greenlanders in a raw haunch of venison.

He would fain have gone further, but, on taking the state of the party into consideration, he resolved that it would be prudent to return. Though each had taken two pairs of Eskimo boots with him, they were now nearly barefooted; and the girl, having lost her tools, was unable to mend the dilapidated footgear.

The mountains they saw were doubtless those of the east coast. The nearest lay N.E. or E.N.E., and are smaller than those on the west, if this may be decided from the smaller quantity of snow on their summits. Dalager thought that, so far as a journey to the east coast across the inland ice was concerned, there was nothing to preclude its possibility in the nature of the ground. The fields of ice were not so dangerous or so full of chasms, or these so deep as was supposed in his day, and is still generally believed in Greenland. Some are hollowed out like a valley, and others so narrow that they could easily be leaped over with the aid of their guns, or, not being long, can be avoided by a short circuit. On the other hand, he points out that there are difficulties almost insuperable in the way. No one could carry provisions sufficient for such a journey, even if they could supply themselves on the other side for the return journey, and the cold is intensely severe. On the 7th they got back to the fjord where they had left their kajaks. Then crossed next morning, and arrived at their tents before nightfall.[1]

3. *Kielsen's Journey in 1830.*—O. B. Kielsen was a whale-fishing assistant at Holstenborg[2] in the Inspectorate of South Greenland, situated at the mouth of a large fjord. On the 1st of March, 1830, Kielsen penetrated in from this fjord with three sledges, and only provided with dogs' food for the first two days, as one is always moderately certain to fall in with reindeer in that section. The 3rd of March brought him to the last inhabited Greenland fishing-station at the bottom of the fjord, and from this he ran as straight as he could into the interior over the land. After having passed the night in a cleft in the rocks, he ran the whole of the next day. The land was for the most part rather level and unvaried, and his course lay over small lakes and streams. The ground also became more deeply covered with snow, which made travel more difficult, and led to a corresponding scarcity of reindeer and fuel. The 5th of March was devoted to reindeer-hunting for selves and dogs, and

[1] David Cranz's 'History of Greenland, &c.' (English translation, 1820), vol. i. p. 18; and Hans Egede Saabye's 'Bruchstücke eines Tagebuches, gehalten in Grönland in 1770 bis 1778 aus dem Dänischen übersetzt von G. Fries' (Hamburg, 1817).

[2] According to Inglefield, in lat. 66° 56' 46" N., long. 53° 42' W. Bonde, however, gives it as 66° 56' N., and 53° 42' W.; while Ulrich, of the Danish navy, makes it 66° 56' 16" N., 53° 40' 37" W.

two were killed. At the same time from a high point he could see the inland ice. The 6th of March saw them up betimes in the morning, and by midday they came to a considerable extended plain. Here the land sloped inwards, and now they saw at their feet the huge extended mass of the great interior ice. They now quickly ran over small hills, lakes, and streams, until they came to a moderately large lake at the end of the inland ice, which was the limit of their journey. After an attempt to climb the ice, Kielsen returned, and had a most troublesome journey. When he reached the fjord, he found that its frozen surface had broken up, so that he had to go overland to the colony, which he reached on the 9th of March, after having gone into the interior on this journey 80 miles in a straight line from Holstenborg.[1]

4. *Hayes' Journey in* 1860.—The voyage of Dr. I. I. Hayes in the American schooner *United States*, to Smith Sound, in 1860-61, has been so frequently referred to in the public journals that its objects and ends must be familiar to most of my readers. One of the minor excursions which he took, while his vessel lay in winter-quarters, was to the interior of the country, and deserves in this place a notice, as not only one of the most successful of these attempts to penetrate the inland ice, but as also the most northerly of them.

The particular off-shoot of the great interior *mer de glace* (for he was never on the real inland ice, which differs considerably from that which he travelled over) on which he broke ground was that one named by Dr. Kane "My Brother John's Glacier," in Port Foulke, lat. 78° 17' 41" N., long. 72° 30' 57" W. On the advice of his dog-driver, Jensen, he dispensed with dog-sledges; though he afterwards regretted this, as he had reason to believe that on some part of the journey they would have been available. Everybody was keen to go, as it was one of their first attempts at exploration after they got into winter-quarters; but Hayes selected as companions Mr. Knorr, John McDonald, Harvey Heywood, Christian Petersen (a Dane), and the Greenland Eskimo Peter. They set out on the 22nd of October with one sledge and a small canvas tent, two buffalo-skins for bedding, a cooking-lamp, provisions for eight days, and an extra pair of fur stockings, a tea-cup and an iron spoon for each man. Their first camp was at the foot of the glacier, when the temperature was 11° Fahr. The second day they got to the top of the glacier, with hard work and some trifling accidents, one of which threatened to be rather serious, Dr. Hayes having, owing to

---

[1] Rink's 'Grönland Geograph. og Statistisk beskrevet,' Band ii. pp. 97-99.

the party not being roped together, fallen through a *crevasse*; and, as none of the party seemed to have the slightest experience of glacier travel, the wonder was that more mishaps did not occur. The ice was at first rough and broken, and almost free from snow. As they penetrated further in, the surface of the glacier became smoother, the great inequality nearer the edge was probably owing to the inequality of the surface over which it spread itself. After journeying for about 5 miles, they pitched their tent on the ice, and slept soundly, though the temperature was several degrees lower than what it was the night before. On the following day they travelled 30 miles, and the ascent, which during the last march had been an angle of about 6°, diminished to about one-third of that angle of observation; and from a surface of hard ice they had come upon a plateau of compacted snow, through which no true ice could be got by digging down to the depth of three feet. At that depth, however, the snow assumed a more gelid condition, and, though not actually ice, they could not penetrate into it further without great difficulty. The snow was covered with a crust which the foot broke at every step, making the travelling very laborious. About 25 miles were made the following day, the track being much the same character, and at about the same elevation. The temperature had now fallen to 30° below zero (of Fahrenheit), and a fierce gale meeting them in the face, drove them to the shelter of their tent, and, after resting for a few hours, compelled them to return, though Dr. Hayes had intended proceeding one day further when he first set out. The temperature was now 34° below zero during the night, though at Port Foulkes, during their absence, it was 22° higher. All of them were more or less frost-bitten, and one of the party seemed likely to give in altogether. The cold was so intense that all of them had to quit the shelter of the tent and run about on the ice to save themselves from getting benumbed. They were now at an altitude of 5000 feet above the sea, 70 miles [1] from the coast, in the midst of a vast frozen Sahara, immeasurable to human eye. Neither hill nor dale was anywhere in view. They had completely sunk the strip of land which lies between the *mer de glace* and the sea, and no object met the eye but their feeble tent, which bent to the storm. "Fitful clouds swept over the face of the full-orbed moon, which, descending towards the horizon, glimmered through the drifting snow that whirled out of the illimitable

---

[1] In the American 'Proc. Philosoph. Soc.,' Dec., 1861, and 'Proc. Royal Geogr. Soc.,' vol. ix. p. 186, Dr. Hayes mentioned the distance which he penetrated into the interior as *fifty* miles. With every respect to him, I think that he has over-estimated the distances travelled by his party on the glacier.

distance, and scudded over the icy plains ; to the eye in undulating lines of downy softness—to the flesh in showers of piercing darts."

The storm now caused them to run for life to an elevation of 3000 feet lower before they stopped, when the wind was less severe, and the temperature 12° higher. Next day they reached Port Foulke without any serious accident, the latter part of their journey being wholly by moonlight. Hayes' journey was undertaken at much too late a period of the year ; but still, so far as it went, it was conducted with all the *esprit* and reckless courage in which his nation has never been wanting, either in battle or in geographical exploration, which demands bravery of a calmer and more enduring description. "My Brother John's Glacier" projects into a valley, about 2 miles from the coast, towards which it is gradually approaching. Hayes' measurements show that it is moving seaward at a very rapid rate, viz. 94 feet in 8 months. This will, however, vary according to the season, the nature of the ground traversed, and other mechanical and physical causes.

5. *Rae's attempted Journey in* 1860.—While Hayes was struggling into winter-quarters in Smith Sound, an English surveying steamer, under the command of Captain Allen Young, was searching the South Greenland fjords, in connection with a projected Atlantic telegraph-cable to be laid *vià* Iceland and Greenland. This project has long ago passed into the limbo of forgotten schemes, now that the Altantic is traversed by two submarine cables, but during this survey (in the *Fox*) an attempt was made to penetrate the interior of Greenland : attached to the expedition and in charge of the land party was Dr. John Rae—already most deservedly famous as an Arctic explorer. The expedition reached Fredrikshaab from Iceland on the 2nd of October, and, on the 24th, while the fjord of Igalliko was being sounded, Dr. Rae considered that a short journey should be made to the interior of the country for the purpose of ascertaining the practicability of travelling over it. The use of one seaman and a whale-boat was obtained from Captain Young to enable the party to return from the head of the fjord to Julianehaab. Four Eskimo women—who in South Greenland are commonly engaged in such labour—were engaged as rowers. They never reached the inland ice ; for, after travelling through a miry and boulder-covered valley 16 miles in from the head of the fjord, a heavy fall of snow stopped further travel, and they returned, after an absence of four days, to their boat—not, however, before the fjord was frozen up for several miles—and with much difficulty they reached the *Fox*.

Mr. Whymper's Expedition in 1867.—Towards the end of July 1867, the present writer, in company with Mr. Edward Whymper (who most carefully planned the trip and made every arrangement), Mr. Anthon P. Tegnér, Mr. Jens Fleischer, and Amac, a Greenland Eskimo (since deceased), made an attempt to penetrate this icy waste with dog-sledges. The season was too late, and our attempt was impeded by various circumstances. Accordingly we only were enabled to proceed for a short distance, when, by the breaking down of our sledges, we were forced to return. Even had this been the place for it, any detailed account of this attempt would take up too much space. The general results obtained by it I have already given.

7. *Visits of Rink and others to the Inland Ice.*—The journeys or attempts which I have recorded at some length form the chief attempts which, as far as I can learn, have been made to penetrate the interior of Greenland, or which have been recorded. Possibly there may have been others, though, from the well-known dislike of the Eskimo to travel over the interior ice, and the absence of any motive for enterprise in that direction on the part of the Danish officers in charge of the government and trade of Greenland, I think that it is hardly likely that there have been many other attempts, and my friend, Dr. Rink, the most distinguished authority on all matters Greenlandic, and for so many years Royal Inspector of South Greenland, whom I consulted on the subject, agrees with me. However, in addition to those I have recorded at length, there are one or two of which I have no notes, or very brief ones, to mention. Dr. Rink himself, who has been close to and has partly viewed and delineated the margin of the inland ice in many difficult places from 60° to 70° N.E., has also ascended the ice itself, namely, at Tessiurssak, near Claushavn, in May, 1851 ; but only spent some hours in walking upon it and in examining its surface, without the intention of trying any inland excursion.

I am also informed by Dr. Rink that a Danish gentleman who visited Greenland in 1862, for the purpose of magnetical observations, has walked several miles over this inland ice near Pakitsok.

The natives are generally reindeer-hunting close to the margin of the ice, and sometimes cross parts of it. A native gives an account of this in the Greenland Journal, 'Atuagagdliutit' of 1864, in the Eskimo language. As, for instance, he says (' Atuag.' p. 451), mentioning the localities from 64° to 65° N.: "On some of the hunting-grounds there are dangers to be encountered, namely, as follows :—The rivers issuing from the ice are very muddy, also when walking over the ice (it presents itself) very fissured, the

crevasses in which cannot be crossed, but must be gone around, are
tremendously deep. If somebody should fall into them he could
never be saved. The reindeer-hunters used to come there. The
land ice enlarges rapidly," &c.

The late Mr. Olrik, so many years inspector of North Greenland
and director of the Greenland trade in Copenhagen, and his
brother-in-law and predecessor, the well-known conchologist—
Inspector Möller—also visited the inland ice. In all likelihood,
the feat of exploring the interior will be again attempted this
summer by an eminent Arctic and Alpine explorer.

8. *Nordenskjöld's and Berggren's Journey in 1870.*[1]—The account of
this interesting attempt I give in the leader's words. It is interesting not only as being the most successful one ever made on the
inland ice, but in the fact that it was conducted by a very experienced Arctic explorer, and by men of science so eminent and
accomplished as Professor Nordenskjöld and Dr. Berggren—a well-
known botanist, lately Assistant-Professor in the University of
Lund, and now engaged in botanical travels in New Zealand :—

" If the inland ice were not in motion, it is clear that its surface
would be as even and unbroken as that of a sand-field. But this,
as is known, is not the case. The inland ice is in constant motion,
advancing slowly but with different velocity in different places,
towards the sea, into which it passes, on the west coast of Greenland,
through eight or ten large and a great many small ice-streams.
[For a description of these see p. 38.] This movement of the ice
gives rise in its turn to huge chasms and clefts, the almost bottomless depth of which close the traveller's way. It is natural that
these clefts should occur chiefly where the movement of the ice is
most rapid, that is to say, in the neighbourhood of the great ice-
streams; but that, on the other hand, at a greater distance from
these the ground will be found more free from cracks. On this
account I determined to begin our wanderings on the ice at a point
as far distant as possible from the real ice-fjords. I should have
preferred one of the deep 'ström-fjords' (stream-fjords) for this
purpose; but as other business, intended to be carried out during
the short summer, did not permit a journey, per boat, so far
southward, I selected instead for my object the northern arm of
Auleitsivikfjord, which is situated 60 miles south of the ice-fjord

---

[1] From a translation of his ' Redogörelse för en Expedition till Grönland år
1870,' in the 'Geological Magazine' (edited by Henry Woodward, F.R.S.), 1872
(vol. ix.). pp. 303-306, 355-362. The passages within brackets are mine, and
here and there I have ventured to make some slight emendations on the translation (apparently by the learned traveller himself) when such was obviously
required, but in no case have I in any way altered his meaning.

at Jakobzhavn, and 240 north of that of Godthaab. The inland ice,
it is true, even in Auleitsivik Fjord, reaches to the bottom of the
fjord; but it only forms there a perpendicular glacier, very similar
to the glaciers at King's Bay, in Spitzbergen, but not any real ice-
stream. There was, accordingly, reason to expect that such fissures
and chasms as might here occur would be on a smaller scale.

On the 17th of July, in the afternoon, our tent was pitched on
the shore north of the steep precipitous edge of the inland ice at
Auleitsivikfjord. After having employed the 18th in preparations
and a few slight reconnoitrings, we entered on our wanderings
inward on the 19th. We set out early in the morning, and first
rowed to a little bay situated in the neighbourhood of the spot
occupied by our tent, into which several clayey rivers had their
embouchures. Here the land assumed a character varied by hill
and dale, and further inward was bounded by an ice-wall somewhat
perpendicular and sometimes rounded, covered with a thin layer
of earth and stones near the edge, only a couple of hundred feet high,
but then rising at first rapidly, afterwards more slowly, to a height
of several hundred feet. In most places this wall could not possibly
be scaled; we, however, soon succeeded in finding a place where it
was cut through by a small cleft, sufficiently deep to afford a possibility of climbing up, with the means at our disposal—a sledge—
which at need might be used as a ladder, and a line, originally
100 fathoms long, but which, proving too heavy a burden, had,
before our arrival at the first resting-place, been reduced one-half.
All of us, with the exception of our old and lame boatman, assisted
in the by no means easy work of bringing over mountain, hill and
dale, the apparatus of the ice-expedition to this spot, and after our
dinner's rest, a little further up the ice-wall. Here [as usual] our
followers left us; only Dr. Berggren, I, and two Greenlanders (Isak
and Sisarniak) were to proceed further. We immediately commenced our march, but did not get very far that day. The inland
ice differs from ordinary glaciers by, among other things, the almost
total absence of moraine formations. The collection of earth, gravel,
and stone, with which the ice on the landward edge is covered, are,
in fact, so inconsiderable in comparison with the moraines of even
very small glaciers that they scarcely deserve mention, and no
longer newly-formed ridges of gravel, running parallel with the
edge of the glacier, are to be met with, at least in the tract visited
by us. The landward border of the inward ice is, however, dark-
ened, we can scarcely say covered, with earth, and sprinkled with small
sharp stones. Here the ice is tolerably smooth, though furrowed
by deep clefts at right angles to the border, such as that made use

of by us to climb up. But in order not immediately to terrify the
Greenlanders by choosing the way over the frightful and dangerous
clefts, we determined to abandon this comparatively smooth ground,
and at first take a southerly direction parallel with the chasms, and
afterwards turn to the east. We gained our object by avoiding the
chasm, but fell in instead with extremely rough ice. We now understood what the Greenlanders meant when they endeavoured to dissuade us from the journey on the ice, by sometimes lifting their
hands over their heads, sometimes sinking them down to the ground,
accompanied by to us an unintelligible talk. They meant by this to
describe the collection of closely-heaped pyramids and ridges of ice
over which we had now to walk. The inequalities of the ice were,
it is true, seldom more than 40 feet high, with an inclination of
25° to 30°. But one does not get on very fast when one has continually to drag a heavily-laden sledge up so irregular an acclivity,
and immediately after to endeavour to get down uninjured, at the
risk of getting one's legs broken, when occasionally losing one's
footing on the here often very slippery ice, in attempting to moderate the speed of the downward-rushing sledge. Had we used an
ordinary sledge, it would have been immediately broken to pieces;
but as the component parts of our sledge were not nailed, but tied
together, it held together at least for some hours.

Already the next day we perceived the impossibility, under such
circumstances, of dragging with us the thirty days' provisions with
which we had furnished ourselves, especially as it was evident that,
if we wished to proceed further, we must transform ourselves from
draught to pack horses. We, therefore, determined to leave the
sledge and part of the provisions, take the rest on our shoulders, and
proceed on foot. We got on quicker, though for a sufficiently long
time over ground as bad as before. The ice became gradually
smoother, and was broken by large bottomless chasms, which one
must either jump with a heavy load on one's back—in which case
woe to him who made a false step—or else make a long circuit to
avoid. After two hours' wandering the region of clefts was passed.
We, however, in the course of our journey, very frequently met with
portions of similar ground, though none of any very great extent.
We were now at a height of more than 800 feet above the level of
the sea. Further inward the surface of the ice, except the occa-
sionally-recurring cleft, resembled that of a stony sea-midden, bound
in fetters by the cold. The rise upwards was still quite perceptible,
though frequently interrupted by shallow valleys, the centres of
which were occupied by several lakes or ponds, with no apparent
outlet, though they received water from innumerable rivers running

along the sides of the excavation. These rivers presented in many places not so dangerous, though quite as time-wasting, a hindrance to our progress as the clefts—with this difference, however, that they did not so often occur; but the circuits to avoid them were so much the longer. During the whole of our journey on the ice we constantly enjoyed fine weather; frequently there was not a single cloud visible in the whole sky. The warmth was to us, clad as we were, sensible; higher up, in the shade, as much as 7° or 8° Centigrade [19·4° or 17·6° Fahr.], but in the sun 25° to 30° Cent. [77 to 86° Fahr.]. After sunset[1] the water-pools froze, and the night was very cold; we had no tent with us, and, although our party consisted of four men, only two ordinary sleeping-sacks. These were open at both ends, so that two persons could, though with great difficulty, with their feet opposite to each other, squeeze themselves into one sack. With rough ice for a substratum, the bed was thus so uncomfortable that, after a few hours' sleep, one was awakened by a cramp in one's closely-contracted limbs; and, as there was only a thin tarpaulin between the ice and the sleeping-sack, the bed was extremely cold to the side resting on the ice, which the Greenlanders, who turned back before us, described to Dr. Nordström [one of Professor Nordenskjöld's party in Greenland] by shivering and shaking throughout their whole bodies. Our nights' rests were, therefore, seldom long; but our midday rest, during which we could bask in a glorious warm sun-bath, was taken on a proportionately more copious scale, whereby I was enabled to take observations for both altitude and longitude.

On the surface of the inland ice we do not meet with any stones at a distance of more than a cable's length from the border; but we find everywhere, instead, vertical cylindrical holes, of a foot or two deep, and from a couple of lines to a couple of feet in section, so close one to another that one might in vain seek between them room for one's foot, much less for a sleeping-sack. We had always a system of ice-pipes of this kind as a substratum when we rested for the night; and it often happened, in the morning, that the warmth of our bodies had melted so much of the ice, that one's sleeping sack touched the water wherewith the holes were always nearly full. But, as a compensation, wherever we rested, we had only to stretch out our hands to obtain the very finest water to drink. The holes in the ice filled with water are in no way connected with each other, and at the bottom of them we found everywhere, not only near the border, but in the most distant parts of the inland ice visited by us, a layer, some few millimetres thick,

---

[1] The reader must, however, remember that at that season there was continuous daylight throughout the twenty-four hours.—[ED.]

C

of grey powder, often conglomerated into small round balls of loose consistency. Under the microscope, the principal substance of this remarkable powder appeared to consist of white angular transparent crystals. We could also observe remains of vegetable fragments; yellow, imperfectly translucent particles, with, as it appeared, evident surfaces of cleavage (felspar), green crystals (augite) and black opaque grains, which were attracted by the magnet. The quantity of these foreign components is, however, so inconsiderable, that the whole mass may be looked upon as one homogeneous substance. An analysis, by Mr. G. Lindström, of this fine glacial sand gave :—

Silicic acid	62·25
Alumina	14·93
Sesquioxide of iron	0·74
Protoxide	4·64
Protoxide of manganese	0·07
Lime	5·09
Magnesia	3·00
Potassa	2·02
Soda	4·01
Phosphoric acid	0·11
Chlorine	0·06
Water, organic substance (100° to red-heat)	2·86
Hygroscopic water (15° to 100°)	0·34
	100·12

Hardness inconsiderable, crystallization probably monoclinic. The substance is not a clay, but a sandy trachytic mineral, of a composition (especially as regards soda) which indicates that it does not originate in the granite region of Greenland. Its origin appears therefore to me very enigmatical. Does it come from the basalt region? or from the supposed volcanic tracts in the interior of Greenland? or is it of meteoric origin? The octahedrally-crystallised magnetic particles do not contain any traces of nickel. As the principal ingredient corresponds to a determinate chemical formula, it would perhaps be desirable to enter it under a separate class in the register of science, and for that purpose I propose for this substance the name of *Kryokonite* (from κρούς and κόνις).

When I persuaded our botanist, Dr. Berggren, to accompany me in the journey over the ice, we joked with him on the singularity of a botanist making an excursion into a tract, perhaps the only one in the world, that was a perfect desert as concerns botany. This expectation was, however, not confirmed. Dr. Berggren's quick eye soon discovered, partly in the surface of the ice, partly in the above-mentioned powder, a brown polycellular alga, which, little as it is, together with the powder and certain other microscopic organisms by which it is accompanied, is the most dangerous enemy to the mass of ice, so many thousand feet in height, and hundreds of miles in extent. The dark mass absorbs a far greater amount of the sun's rays of heat than the white ice, and thus produces over its whole surface deep holes which greatly promote the process of melting. The same plant has no doubt played the same part in our country, and we have to thank it, perhaps, that the deserts of ice which formerly covered the whole of northern Europe and America, have now given place to shady woods and undulating corn-fields. Of course a great deal of the grey powder is carried down in the rivers, and the blue ice at the bottom of them is not unfrequently concealed by a dark dust. How rich this mass is in organic matter is proved by the circumstance, amongst others, that the quantity of organic in it was sufficient to bring a large collection of the grey powder, which had been carried away to a distant part of the ice by sundry now dried-up glacier streams, into so strong a process of fermentation or putrefaction, that the mass, even at a great distance, emitted a most disagreeable smell, like' that of butyric acid." Dr. Berggren has described these organisms in the ' Öfv. Kongl. Vet.-Akademiens Förh.' for 1871, p. 293, under the name of *Ancylonema Nordenskiöldii* Berggr. *Protococcus nivalis* is also common, as well as *P. vulgaris* and *Scytomena gracilis*.

"At our midday rest on the 21st we had reached lat. 68° 21' and 36' long. east of the place where our tent was pitched, and a height of 1400 feet above the level of the sea. Later in the day, at our afternoon rest, the Greenlanders take to take off their boots and examine their little thin feet—a serious indication, as we soon perceived. Isak presently informed us, in broken Danish, that he and his companions now considered it time to return. All attempts to persuade them to accompany us a little farther failed, and we had, therefore, no other alternative than to let them return, and continue our excursion without them. We took up our night's quarters here. The provisions were divided. The Greenlanders, considering that they might perhaps not be able to find our first depôt, were allowed to take as much as was necessary to enable them to reach the tent. We took out cold provisions for five days. The remainder, together with the excellent photogen portable kitchen, which we had hitherto carried with us, were laid up in a depôt in the neighbourhood, on which a piece of tarpaulin was stretched upon sticks, that we might be able to find the place on our return, which, however, we did not succeed in doing, though we must have passed in its immediate vicinity. After these preparations for a parting, Dr. Berggren and I proceeded alone further inward. The Greelanders turned back. At first we passed one of the above-mentioned extensive bowl-formed excavations in the ice-plain, which is here furrowed by innumerable rivers, which often obliged us to make long circuits; and when to avoid this we endeavoured to make our way along the margin of the valleys, we came instead upon a tract where the ice-plain was cloven by long, deep, parallel clefts, running true N.N.E. to S.S.W., quite as difficult to get over as the rivers, but far more dangerous. Our progress was accordingly but slow. At twelve o'clock on the 22nd we halted in glorious, warm sunny weather to make a geographical determination ; we were now at a height of 2000 feet, in lat. 68° 22′ and in a long. of 56′ of arc east of the position of our tent at the fjord. During the whole of our excursion on the ice we had seen no other animals than a couple of ravens, which on the morning of the 22nd, at the moment of our separation, flew over our heads. At first, however, there appeared in many places on the ice remnants of ptarmigans, which seemed to indicate that these birds visit these desert tracts in by no means inconsiderable flocks. Everything else around was lifeless. Nevertheless, silence by no means reigned here ; on bending down the ear to the ice, one could hear on every side a peculiar subterranean hum, proceeding from rivers flowing within the ice, and occasionally a loud single report like that of a cannon gave notice of the formation of a new glacier cleft.

"After taking the observations, we proceeded over comparatively better ground. Later in the afternoon we saw, at some distance from us, a well-defined pillar of mist, which, when we approached it, appeared to rise from a bottomless abyss, into which a mighty glacier-river fell. The vast roaring water-mass had bored for itself a vertical hole, probably all the way down to the rock, situated certainly more than 2000 feet beneath, on which the glacier rested. The following day (the 23rd) we rested in lat. 68° 22′, and 76′ of arc longitude east from the position of our starting-point at Auleitsivik. The provisions we had taken with us were, however, now so far exhausted, that we were obliged to think of returning. We determined, nevertheless, first to endeavour to reach an ice-hill, visible on the plain to the east, from which we hoped to obtain an extensive view : and, in order to arrive there as quickly as possible, we left the scanty remains of our provisions and our sleeping-sack at the spot where we had passed the night, taking careful notice of the ice-rocks around, and thus proceeded by forced march, without incumbrance.

" The ice-hill was considerably farther off than we had supposed. The walk to it was richly rewarded by an uncommonly extensive view, which showed us that the inland ice continued constantly to rise towards the interior, so that the horizon towards the east, north, and south, was terminated by an ice-border almost as smooth as that of the ocean. A journey further (if one were in a condition to employ weeks for the purpose—which want of time and provisions rendered impossible to us) could, therefore, evidently furnish no other information concerning the nature of the ice than that which we had already obtained; and even if want of provisions had not obliged us to return, we should hardly have considered it worth while to add a few days' marches to our journey. Our turning-point was situated at a height of 2200 feet above the level of the sea, and about 83' of longitude, or 30 miles west of the extremity of the northern arm of Auleitsivik Fjord. On departing from the spot where we had left our provisions and sleeping-sack, we had, as we supposed, taken careful notice of the situation: nevertheless, we were nearly obliged to abandon our search as vain—an example which shews how extremely difficult, without lofty signals, we find objects again on a slightly undulating surface everywhere similar, like that formed by the inland ice. When, after anxiously searching in every direction, we at length found our resting-place, we ate our dinner with an excellent appetite, made some further reductions in our load, and then set off with all haste to the boat, which we reached late in the evening of the 25th.

" At a short distance from our turning-point we came to a copious, deep, and broad river, flowing rapidly between its blue banks of ice, which were here not discoloured by any gravel, and which could not be crossed without a bridge. As it cut off our return, we were, at first, somewhat disconcerted: but we soon concluded that, as on our journey out we had not passed any stream of such large dimensions, it must, at no great distance, disappear under the ice. We therefore proceeded along its banks in the direction of the current, and, before long, a distant roar indicated that our conjecture was right. The whole immense mass of water here rushed down a perpendicular cleft into the depths below. We observed another smaller, but, nevertheless, very remarkable waterfall the next day, while examining, after our mid-day rest, the neighbourhood around us with a telescope. We saw, in fact, a pillar of steam rising from the ice at some distance from our resting-place, and, as the spot was not far out of our way, we steered our course by it, in the hope of meeting—judging from the height of the misty

pillar—a waterfall still greater than that just described. We were
mistaken : only a smaller, though, nevertheless, tolerably copious,
river rushed down from the azure cliffs, to a depth from which no
splashes rebounded to the mouth of the fall : but there arose instead,
from another smaller hole in the ice, in the immediate vicinity, an
intermittent jet of water mixed with air, which, carried hither and
thither by the wind, wetted the surrounding cliffs with its spray.
We had, then, here, in the midst of the desert of inland ice, a
fountain, as far as we could judge from descriptions, very like the
Geysers, which in Iceland are produced by volcanic heat.

"In order, if possible, to avoid the district of the rocks, which, on
our journey out had required so much patience and exertion, we
had, on returning, chosen a more northerly route, intending to
endeavour to descend from the ice-ridge up on the slip of ice-free
land which lies between the inland ice and Disco Bay. The ice
was here, with the exception of a few ice-hillocks of a few feet
high, in most places as even as a floor, but often crossed by very
large and dangerous clefts, and we were so fortunate as immediately
to hit upon a place where the inclination towards the land was
inconsiderable, so that one might have driven up a four-in-hand.
The remainder of the way along the land was harder, partly on
account of the very uneven nature of the ground, and partly on
account of the numerous glacier-streams which we had to wade
through, with the water far above our boots. At last, at a little
distance from the tent, we came to a glacier-stream, full of muddy
water, so large that, after several failures, we were obliged to
abandon the hope of finding a fordable place. We were, therefore,
obliged to climb high up again on the shining ice, so as to be able
to find our way down again further on, after passing the river;
but the descent on this occasion was more difficult than before."

9. *What is the Interior of Greenland ?*—It may seem a paradox when
I say that so far as we can draw any conclusions from the observations on the short journeys into the country described in the foregoing pages, *Greenland has no Interior !* At least if we look upon
its interior in the light of something else than ice and snow. Solid
land or rock there is none now to be seen. All that we know of it
shows it to be "a waste and weary land where no man comes, or
hath come, since the making of the world." The country seems
only a circlet of islands separated from one another by deep fjords
or straits, and bound together on the landward side by the great
ice-covering which overlies the whole interior, and which is pouring out its overflow into the sea in the shape of glaciers and icebergs. No doubt, under this ice there lies land, just as it lies under

the sea; but nowadays none can be seen, and as an insulating medium it might as well be water. Cross over that surrounding circlet of outskirting island, and we ascend to a plateau where nought can be seen but ice. No fragment of stone is there—no trace of vegetation, except a trace here and there of the red snow-plant—not a sight or sound of moving thing, nothing but hard glacier ice stretching north and south—westward after you have lost sight of the land you have crossed over, and eastward as far as the eye can see. The mountains which Dalager saw in South Greenland to the eastward were in all likelihood those of the East Coast, and not interior mountains, for wherever else it has been penetrated into, nothing but ice can be seen on the distant eastern horizon. How deep this ice overlies the country it is impossible to say; in some places, I doubt not, many thousand feet. As I have already, in the section on the Glaciers and Ice of Greenland, described the nature of this glacial covering at some length, it is not necessary for me to go into a description of it in this place. I see no reason to doubt that it continues throughout the whole country, except where fjords may indent it, and even then, in many cases, it is increasing —it is filling up these fjords. Dr. Rink has also discussed this subject,[1] in a paper in the Danish 'Tidsskrift for populair Fremstilling af Naturvidenskab' for October, 1870, as well as in a recently published brochure.[2]

10. *Are there any Mountains in the Interior?*—From what I have just said, it will be apparent that there are none of any extent. Whatever there may have been formerly are now overlaid by an ice-covering, viz., by the glacial cap forming, by the immense fall of snow and the little evaporation in the cold interior, much more rapidly that it can be discharged in the shape of icebergs. There are no iceberg "streams" on the east coast of Greenland, and bergs are rare off that coast. As soon as you leave the immediate vicinity of the coast no moraine is seen coming over the inland ice, which

---

[1] " Om Grönlands indland, og muligheden af at Bereise samme " [On the Interior of Greenland, and the possibility of Exploring the same], No. 9 of ' Fra Videnskabens Verden.' Copenhagen, 1875.

[2] " The whole interior of the country, indeed," writes Mr. James Geikie, and I quote his conclusions as peculiarly bearing on the subject, " would appear to be buried underneath a great depth of snow and ice, which levels up the valleys, and sweeps over the hills. The few daring men who have tried to penetrate a little way from the coast, describe the scene as desolate in the extreme—far as the eye can reach, nothing save one dead, dreary expanse of white. No living creature frequents this wilderness—neither bird, nor beast, nor insect—not even a solitary moss or lichen can be seen. Over everything broods a silence deep as death, broken only when the roaring storm arises to sweep before it the pitiless blinding snow."—' The Great Ice Age,' p. 56.

would certainly not be the case if the ice sloped from any mountain range or in its tract to the coast touched any land at all. No living creature—animal or plant—appeared on this desolate glacier-field except a trace here and there of the red snow-plant (*Protococcus nivalis, P. vulgaris,* &c.), so common in Alpine and Arctic regions. I find, however, that Dr. Berggren discovered, as already noted, what in our anxiety and other duties we might have omitted to observe—various low forms of vegetable life, chiefly *Diatomaceæ*—though approaching the *Zygonemaceæ* (*Scytonema gracilis,* &c.). These might be expected, as we continually find them in hollows of icebergs (*vide* Sutherland's 'Arctic Voyage with Captain Penny,' and my paper on the discolouration of the sea [1]—the facts in which have been confirmed both by the Germans and Swedes. I am therefore of opinion that the great ice-field slopes from the east to the west coast of Greenland (chiefly),[2] and that any bergs which may be seen on the coast are from local glaciers, or from some unimportant defluent of the great interior ice. Nor do I think a range of mountains at all necessary for the formation of this huge *mer de glace*, for this idea is derived from the Alpine and other mountain ranges where the glacial system is a petty affair compared with that of Greenland. I look upon Greenland and its interior ice-field—to recapitulate what I will have occasion more fully to enter upon when describing the inland ice (p. 34)—in 'the light of a broad-lipped, shallow vessel, but with breaks in the lips here and there, and the glacier like some viscous matter in it. As more is poured in, the viscous matter will run over the edges, naturally taking the line of the chinks as its line of outflow. The broad lips of the vessel, in my homely simile, are the outlying islands or " outskirts ;" the viscous matter in the vessel the inland ice, the additional matter continually being poured in the enormous snow covering, which, winter after winter, for seven or eight months in the year, falls almost continuously on it ; and the chinks or breaks in the vessel are the fjords or valleys down which the glaciers, representing the outflowing viscous matter, empty the surplus of the vessel. In other words, the ice flows out in glaciers—overflows the land, in fact, down the valleys and fjords of Greenland—by force of the superincumbent weight of snow, just as does the grain on the floor of a barn when another sackful is emptied on the top of the mound already on the floor. The want of much slope, therefore, in the country, and the absence of any great mountain range,

---

[1] 'Trans. Botanical Society Edin.,' vol. ix.
[2] Quart. Jour. Geol. Soc. Lond., 1871,' pp. 671-701.

are of little moment to the movement of this (or any other great mass of land-ice) *provided ice have snow enough.* In the Appendix to Lyell's 'Antiquity of Man,' p. 508, it is stated that Professor Otto Torrell, of Lund, Director of the Geological Survey of Sweden, from Mount Karsok in the Noursak Peninsula, North Greenland, saw the inland ice with some "abrupt mountains standing up here and there," and that, at Upernavik, Rink saw moraines on the ice. I am inclined to believe that these were only local, and the mountains were not in the midst of the inland ice proper, but only part of those on the outskirting land. No moraine comes over it from the south.

11. *What is Greenland?*—Greenland, as it appears on our maps, is a huge wedge of land hanging down from the North Pole. Add to this the exaggerated proportions which Mercator's projection gives to it, and the ranges of interior mountains which imaginative geographers now and then portray in its interior, and we are all sufficiently familiar with its outline. It is now more than half a century ago since Giesecke,[1] who had long resided in the country, expressed his opinion that it was merely a collection of islands bound together by ice; and from what I have said, further research has not invalidated, though it may have supported and extended his views. Dr. Petermann considered that it might extend in a more or less unbroken line to Wrangell's Land, north of Behring Strait. With the views of Giesecke I am inclined to concur. That the idea of Kane and Hayes, that it ends in an "open Polar Sea," is unsupported and unreasonable, there can, I think, be little doubt, and the idea is not now coincided in by many whose opinions on such a matter can be received as of much moment. That it is a collection of islands bound together by the inland ice and its outpouring glaciers I have already ventured to state my belief as being a well-observed fact, and that, in a collection of broken islands, it extends throughout the Arctic Polar basin perhaps on to Wrangell's Land is, I further believe, not at all improbable. Shortly before writing these notes I read the admirable papers of Lieutenant Payer on Kaiser Franz Joseph Fjord;[2] and while admitting that this and many other east-coast fjords may penetrate the land for great distances, I do not think that his views tend materially to alter the doctrine I have stated. It was long a belief that some of the west-

---

[1] Appendix to Scoresby's 'Voyage to the Northern Whale Fishery,' p. 467, and Scoresby, *ibid.*, p. 327.

[2] 'Geogr. Mitt., 1871,' Heft. iv. and v. This is supposed to stretch far in from the east coast, in lat. 65° (*vide* picture of it by Payer, in Petermann's 'Geog. Mitth.,' 1871, and in the 'Leisure Hour' for Oct. 1871).

coast fjords—particularly those about Omenak Fjord and Disco
Bay—cut Greenland in two (see p. 42), and the Eskimo to this
day have traditions of timber drifting out, and even of men coming
through these fjords from the east coast. But whether this was
so or not in former times, we know this is not so now, and as all
of the west Greenland fjords are known as to their termination,
there need be little or no doubt as to the fact of Franz Joseph Fjord
not now reaching through to the west coast. Though the exact
heads of some of these fjords have not been reached, it is known
that they are terminated by the ice face of a glacier. So that,
though there may not be now water communication between the
east and west coast, it is just possible that at one time, before the
spread of the inland ice choked up these fjords (as we know it has
done Jakobshavn ice fjord and others within the memory of man),
it may have been so in former times; and even yet there may be
no land shutting off the one end of the fjord from the other. The
Germans did not see the inland ice. That means nothing more
than that they did not penetrate far enough to pass over the outskirting land.

12. *Can Greenland be crossed?*—It may, I think, over the smooth,
snow-covered inland ice at certain seasons of the year, say in
May, when it is tolerably mild, and the whole summer is before
us, and the snow has not yet melted off the ice. Later in the
season the snow melts off the ice, and, as happened in our
case, travel was impossible with sledges. Later, again, as when
Dalager and Hayes travelled, the winter is coming on, the nights
are dark, and the cold is intense. After much hardship and with a
fortuitous concourse of favourable circumstances, the country might
be crossed to the east coast, but I do not think the travellers could
return the same way. For even were it possible for them to carry
provisions for themselves and dogs, even allowing them to eat
their spare dogs now and then, it would certainly not be possible
to carry enough for the return journey also, if even the snow covering still remained on the ice. It would be too great a risk to
depend on getting provisions by reindeer-hunting on the east coast,
so that a depôt or a ship would be needed to await them there.
To return down the east coast would be almost as dangerous and
risky as to return across the inland ice. However, in South Greenland, where the continent is narrow, it might be possible to accomplish this. Hitherto I have spoken of a journey from the west *to*
the east coast, because visits to the latter coast are so rare and
difficult, that I had left out of account the chances of any one ever
attempting it there. Still there is a chance of it being done, and

done much more safely and easily from the east than from the west coast. It is even possible that, penetrating the country from Franz Josef or other fjord, and then taking to sledge at a favourable time of the year, that the journey could be performed with comparative ease, for, once arrived at the west coast, there would not be much difficulty in getting succour from the Eskimo or Danish settlements.

I do not despair of its being done; and if judiciously gone about, I do not think the risks are greater than the problem to be solved.

3. GREENLAND GLACIERS AND SEA-ICE.[1]

It is difficult—if not impossible—to describe Greenland glaciers without trenching on subjects of hot and, shall I say, heating controversy. In touching again on the subject of Arctic ice-action and glacial remains in Britain, I am well aware that I am risking the stirring up of a hardly subsided degree of controversy most disquieting to the peace of mind of men unwilling to enter the lists of combatants. Of late years, however, the subject has received new light from the hypothesis, propounded first, I believe, by Agassiz,[2] that Scotland and other portions of the north of Europe were at one time covered with an icy mantle, and that it is to this, and not to the agency of floating ice, that the glacial[3] markings and remains so abundantly scattered over our country are due. More recently still, this theory, at one time so violently opposed, has been brought into almost universal favour by the publication of the fact that Greenland is at this day exactly in the condition in which Agassiz, reasoning on observed facts, hypothetically described North Britain to have been. This new start has been chiefly due to the writings of Dr. H. Rink, of Copenhagen (until recently, and for many years previously, Royal Inspector of South Greenland, and now Director of the Royal Commerce of Greenland), translated in the 'Journal of the Royal Geographical Society,'[4] though the facts were known long previously to his placing them before English geographers in a clear light. Accordingly, thanks

---

[1] This paper is, to a great extent, reprinted from the "Physics of Arctic Ice" ('Quarterly Journal of the Geological Society,' vol. xxvii., 1871, p. 671.
[2] 'Edin. New Phil. Journ.,' vol. xxxiii., p. 217; 'Proc. Geol. Soc., vol. iii., p. 327.
[3] I use the word "Glacial" as expressing all relating to ice, on sea or land; while the word glacier is, of course, used in the ordinary acceptation of the term.
[4] Vol. xxiii. p. 145 (1853); 'Proc. of Soc., vol. vii. p. 76 (1863). It was also described by Dr. Sutherland (from Rink) in Inglefield's 'Summer Search for Sir John Franklin' (1853), Appendix. p. 163.

to the labours of Smith of Jordanhill,[1] Lyell,[2] Chambers,[3] Milne-
Home,[4] Darwin,[5] Fleming,[6] Murchison,[7] Peach,[8] Jamieson,[9] Ramsay,[10]
Thomas Brown,[11] Crosskey,[12] McBain,[13] Howden,[14] Jolly,[15] Archibald
Geikie,[16] James Geikie,[17] and many other geologists, we are in
possession of a body of facts which enable us to reason on the
subject with a degree of certainty which would otherwise have
been impossible. Let us then examine in a concise manner the
subject of the present glaciation of Greenland and other Arctic
countries, and ice-action generally.

Previously to doing so, I may say that I have enjoyed opportunities of studying ice-action in British Columbia, Washington
Territory, Oregon, California, &c., and on the western and eastern
shores of Davis Straits and Baffin Bay—that I have voyaged over
the seas of Spitzbergen and Greenland—that I have passed a whole
summer in the Danish possessions in Greenland, at a post situated
in close proximity to the great ice-fjord Jakobshavn, one of the chief
sources of icebergs in Mid-Greenland—and that, as already mentioned, I was one of those who attempted a journey over this great

[1] 'Quart. Journ. Geol. Soc.,' vol. vi.; 'Memoirs of the Wernerian Natural
History Society,' vol. viii.; and 'Newer Pliocene Geology.'
[2] 'Proc. Geol. Soc.,' vol. iii.; 'Antiquity of Man;' 'Elements' and 'Principles,' &c. &c.
[3] 'Ancient Sea Margins,' and 'Edin. New Phil. Journ.' 1853 and 1855.
[4] 'Coal-fields of Mid-Lothian;' 'Trans. Roy. Soc. Edin.,' vol. xvi.; ibid.
vol. xxv. 1869, &c.   [5] 'Phil. Trans., 1839.'
[6] 'The Geological Deluge, as interpreted by Baron Cuvier and Professor
Buckland, inconsistent with the Testimony of Moses and the Phenomena of
Nature;' 'Lithology of Edinburgh,' &c.
[7] 'Brit. Assoc. Rep.,' vol. xx.; 'Proc. R.G.S.,' vol. vii.; 'Russia in Europe.'
&c. &c.
[8] 'Proceedings of the Royal Physical Society,' Edin. 1861; 'Edin. New Phil.
Journ.,' n. s. vol. ii. &c.
[9] 'Quart. Journ. Geol. Soc.,' vols. xiv. xvi. xviii. xix. and xxiv.
[10] 'Quart. Journ. Geol. Soc.,' vol. xviii.; 'Glaciers of Wales,' &c.
[11] 'Trans. Roy. Soc. Edin.,' vol. xxiv.
[12] 'Trans. Geol. Soc. Glasgow,' vols. ii. and iii.
[13] 'Proc. Roy. Phys. Soc. Edin.' 1859-1862.
[14] 'Proc. Roy. Phys. Soc.,' and 'Trans. Geol. Soc. Edin.,' vol. i.
[15] 'Trans. Geol. Soc. Edin.' vol. i.
[16] 'Scenery of Scotland;' 'Edin. New Phil. Journ.' 1861; 'Trans. Geol. Soc.
Glasgow,' vols. i. iii. &c.
[17] 'Trans. Geol. Soc. Glasgow,' vol. iii.; 'The Great Ice Age' (1874). That
this list by no means exhausts the names of those who by their writings have
advanced the subject, or contains all the papers of those mentioned, is self-
evident. The names of Bald, Imrie, Hall, MacCulloch, Dick-Lauder, Trevelyan,
J. D. and E. Forbes, Hibbert, Maxwell, Prestwich, Maclaren, Craig, Lands-
borough, Mackenzie, Professor Jas. Thomson, Nicol, Cunning, Cleghorn, Smith,
Miller, Hopkins, Brickenden, Bryce, Martin, Hall, Macintosh, Murphy, Lubbock,
the Duke of Argyll, Searles, Wood, jun, Croll, De Rance, and others, are
familiar as having done good service; but I have only referred to the papers
which have come immediately before me.

interior ice-cap. I may, however, mention that in 1867, we were not far enough north, or early enough in Davis Straits, to see anything of the action of sea-ice, and that, though I saw the "inland ice" close at hand for the first time that year, yet I added nothing to the knowledge which my observations during a much more extended voyage along the northern shores of Greenland and the western shores of Davis Straits enabled me to gain as early as 1861. Accordingly many of these descriptions are written almost verbatim from my notes of that date, and the views I now enunciate were formed at that period also. I am, in addition, not ignorant of the remains of the glacial period in Scandinavia and Great Britain, as well as in North America and other countries. Though the facts here narrated will, in almost every case, be wholly derived from my own observation, I wish it to be distinctly understood that I do not present them as any thing new, but solely as the observations and conclusions of an independent student of the subject, and as therefore of some value. If some of the facts here related are already familiar to the reader from other sources, I can only plead that few, if any, of them are yet sufficiently well understood, or received into the commonwealth of knowledge as confirmed facts, not to admit of being repeatedly described by independent observers.

4. GLACIER-SYSTEM OF GREENLAND.

Greenland is in all likelihood a large wedge-shaped island, or series of islands, surrounded by the icy Polar basin on its northern shores, and with Smith Sound, Baffin Bay, Davis Straits, and the Spitzbergen, or Greenland Sea of the Dutch, the "old Greenland Sea" of the English whalers, completing its insularity on its western and eastern sides. The whole of the real de facto land of this great island consists, then, of a circlet of islets, of greater or less extent circling round the coast, and acting as the shores of a great interior mer de glace—a huge inland sea of fresh-water ice, or glacier, which covers the whole extent of the country to an unknown depth. Beneath this icy covering must lie the original bare ice-covered country, at a much lower elevation than the surrounding circlet of islands. These islands are bare, bleak, and more or less mountainous, reaching to about 2000 feet; the snow clears off, leaving room for vegetation to burst out during the short Arctic summer. The breadth of this outskirting land varies, as do the spaces between the different islands. These inlets between the islands constitute the fjords of Greenland, and are the channels through which the overflow of the interior ice discharges itself. It is on these islands, or outskirting land, that the population of Greenland lives, and the Danish trading-posts are built—all the rest of the country, with the exception of this island circlet, being an icy, landless, sea-like waste of glacier, which can be seen here and there peeping out in the distance. On some of the large and more mountainous islands, as might be expected in such a climate, there are small independent glaciers, in many cases coming down to the sea, and there discharging icebergs; but these glaciers are of little importance, and have no connection with the great internal ice-covering of the country. I have called the land circling this interior ice desert "a collection of islands," because though many of them are joined together by glaciers, and only a few are wholly insulated by water, many of them (indeed, the majority) are bounded on their eastern side by this internal inland ice; yet, whether bounded by water or by ice, the boundary is perpetual, and whatever be the insulating medium, they are to all intents and purposes *islands*.

1. *The Interior Ice-field.*—This is well known to the Danes in Greenland by the name of the " inlands iis," and though a familiar subject of talk amongst them from the earliest times, it is only a very few of the " colonists " who have ever reached it. The natives everywhere have a great horror of penetrating into the interior, not only on account of the dangers of ice-travel, but from a superstitious notion that the interior is inhabited by evil spirits in the shape of all sorts of monsters.

Crossing over the comparatively narrow strip of land, the traveller comes to this great inland ice (fig. 1, *a*). If the termination of it is at the sea, its face looks like a great ice wall : indeed the Eskimo called it the *Sermik soak*, which means this exactly. The height of this icy face varies according to the depth of the valley or fjord which it fills. If the valley is shallow the height is low ; if, on the contrary, it is a deep glen, then the sea-face of the glacier in the fjord is lofty. From 1000 to 3000 feet is not uncommon. In such situations the face is always steep, because bergs are continually breaking off from it; and in such situations it is not only dangerous to approach it, on account of the ice falling, or the wave caused by the displacement of the water, but from the great steepness of the face it is rarely possible to get on to it in such situations.[1] In such places Dr. Rink has generally found that it rises by a gradual slope to the general level plateau beyond.[2]

---

[1] The "great glacier" of Humboldt is merely such an exposed glacier-face, though of great extent.

[2] Kane speaks about the "escaladed structure" of the Greenland glacier ('Arctic Explorations' [American ed.], vol. ii. p. 284). This phrase seems to

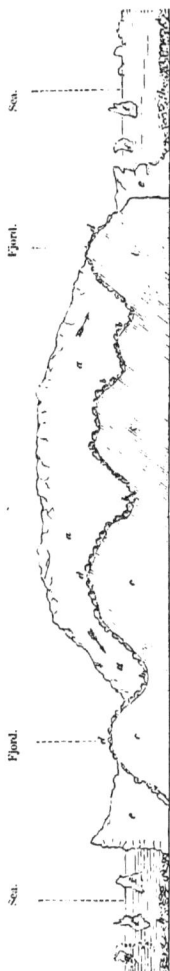

FIG. 1.—DIAGRAMMATIC TRANSVERSE SECTION OF GREENLAND FROM E. TO W. IN ABOUT LAT. 69° N.

Sea.   Fjord.   Fjord.   Sea.

a. The inland ice overlaying the whole interior of the country. b. The "moraine profonde." c. The underlying country, now concealed by the inland ice. d. The present coast (the "outskirts"), covered with old sea-bottom. e. Iceberg in process of formation; i.e. the glacier has protruded into the sea, ground along the bottom, but the buoyancy of the sea has not yet floated off the berg. f. Iceberg floating off. g. Iceberg capsizing and depositing carried moraine on h, the present sea-bottom, composed of boulder clay, angular travelled blocks, &c.

N.B. The glacier commonly does not reach the open sea directly, but enters a fjord, though, for the sake of showing its termination, in the diagram it is portrayed as if reaching the open sea at once. The two arrows show the direction of the outflow of the inland ice (a). I have also portrayed it sloping equally either way. In reality, however, I believe it slopes more to the west than to the east, as the fewer "ice streams" on that coast would seem to show.

However, where it does not reach the sea, it is often possible to climb on it from the land by a gentle slope, or even in some cases to step up on it as it shelves up.   Once fairly on the inland ice a dreary scene meets the view.   Far as the eye can reach to the north and to the south is this same great ice-field, the only thing to relieve the eye being the winding black circuit of the coast-line land or islands before described, here and there infringing in little peninsulas on the ice, there the ice dovetailing in the form of a glacier on the land, and now and then the waters of a deep fjord penetrating into the ice-field, its circuit marked by the black line of coast surrounding it on either side, the eastern generally being the ice-wall of the glacier, the western being the sea.   Travelling a short distance on this interior ice, it seems as if we were travelling on the sea.   The land begins to fade away behind us like the shore receding as we sail out to sea; while far away to the eastward nought can be seen but a dim, clear outline like the horizon bounding our view.   The ice rises by a gentle slope, the gradient being steeper at first, but gradually getting almost imperceptible though real.   In the winter and spring this ice-field must be covered with a deep blanket of snow, and the surface must then be smooth as a glassy lake; but in the summer, by the melting of the snow, it is covered with pools and coursing streams of icy-cold water, which either find their way over the edge, or tumble with a hollow sound through the deep *crevasses* in the ice.   How deep these crevasses go, it is impossible to say, as we could not see to the bottom of them, nor did the sounding-cord reach down except a short way.   The depth of the ice-covering will of course vary; when it lies over a valley it will be deeper, over a mountain-top less.   All we know is, that just now it is almost level throughout, hill and dale making no difference.   However, with such a huge superincumbent mass of ice, the average height of the coast-lying islands is greater than that of the inland ice, and it is only after climbing considerable heights that it can be seen.[1]   Therefore supposing this covering to be removed, I think the country would look like a huge, shallow, oblong vessel with high walls around it.   The surface of the ice is ridged and furrowed after the manner of glaciers generally; and

have arisen from the translator of Dr. Rink's abstract in the 'Journ. Royal Geog. Soc.,' *l. c.*, having mistaken the word "ice-stream" for "ice-steps."   The "ice-steps," or "platform," so universally described by the authors who have followed the translation of Dr. Rink's remarks, have no existence in nature, or in the writings of the eminent geographer mentioned.

[1] In Rink's 'Grönland,' ii. p. 2, are two characteristic views of the appearance of the interior ice seen from such elevations.

this furrowing does not decrease as we go further inland; on the
contrary, as far as our limited means of observation go, it seems to
increase ; so that even were it possible to cross this vast icy-desert
on dog-sledges when the snow is on the ground, I do not think it
would be possible to return, and its exploration would require the
aid of a ship on the other side. On its surface there appears not a
trace of any living thing except a minute alga ; and after leaving
the little outpouring offshoot of a glacier from it, the dreariness of
the scene is not relieved by even the sight of a patch of earth, a
stone, or aught belonging to the world we seem to have left behind.
Once, and only once, during our attempt to explore this waste did
I see a faint red streak, which showed the existence of the red snow-
plant (*Protococcus nivalis*) ; but even this was before the land had
been fairly left. A few traces of other alga were seen by Dr. Berggren, as I have already intimated (pp. 19 and 24). Animal life seems
to have left the vicinity ; and the chilliness of the afternoon breeze,
which regularly blew with piercing bitterness over the ice-wastes,
even caused the Eskimo dogs to couch under the lee of the sledge,
and made us, their masters, draw the fur hoods of our coats higher
about our ears.[1] Whether this ice-field is continuous from north to
south it is not possible in the present state of our knowledge to
decide ; but most likely it is so. Whether its longitudinal range is
continuous is more difficult to decide, though the explorers already
mentioned saw nothing to the eastward to break their view ; so
that, as I shall immediately discuss, there seems every probability
that in Greenland there is one continuous unbroken level field of
ice, swaddling up in its snowy winding-sheet hill and valley, without a single break for upwards of 1200 miles[2] of latitude, and an
average of 400 miles of longitude, or from Cape Farewell to the
upper extremity of Smith Sound, and from the west coast of
Greenland to the east coast of the same country, a stretch of ice-
covered country infinitely greater than ever was demanded hypothetically by Agassiz in support of his glacier-theory.

2. *The Defluents of this Inland Ice-field.*—Are there any ranges of
mountains from the slopes of which this great interior ice descends?
As I have said, we are not in a position to absolutely decide ; but

---

[1] For description of the effects of the ice in limiting animal and vegetable life
vide the author's "Mammalian Fauna of Greenland," 'Proc. Zool. Soc. Lond.
1868,' p. 337 ; and "Florula Discoana," 'Trans. Bot. Soc. Edin.,' vol. ix. p. 440.
[2] Rink, 'Journ. R. G. S.' *l. c.*, says 800 miles ; but throughout his valuable
works he only speaks of the Danish portion of Greenland, of which it professes
solely to be a description. Jamieson and other writers seem to think that it is
only North Greenland that is covered. All the country, north and south, is
equally swathed in ice.

the probabilities are in favour of the negative.⁻ There are no iceberg " streams " on the east coast of Greenland, and bergs are rare off
that coast. If there were many icebergs, the field of floe-ice which
skirts that coast, and which has prevented exploration except in
very open seasons, would soon be broken up by the force with which
the bergs, breaking off from the land, would smash through the ice-
field, and, acting as sails, help, by the aid of the winds, as elsewhere,
to sweep it away. I am therefore of opinion that the great ice-
field slopes from the east to the west coast of Greenland, and that
any bergs which may be seen on that coast are from local glaciers,
or from some unimportant defluent of the great interior ice. Nor
do I think a range of mountains at all necessary for the formation
of this huge *mer de glace;* for this is an idea wholly derived from
the Alpine and other mountain-ranges where the glacier system is
a petty affair compared with that of Greenland. I look upon Greenland and its interior ice-field in the light of a broad-lipped shallow
vessel, but with chinks in the lips here and there, and the glacier,
like viscous matter[1] in it. As more is poured in, the viscous matter
will run over the edges, naturally taking the line of the chinks as
its line of outflow. The broad lips of the vessel, in my homely simile,
are the outlying islands or " outskirts;" the viscous matter in the
vessel the inland ice, the additional matter continually being poured
in in the form of the enormous snow covering, which, winter after
winter, for seven or eight months in the year, falls almost continuously on it; the chinks are the fjords or valleys down which
the glaciers, representing the outflowing viscous matter, empty the
surplus of the vessel. In other words, the ice floats out in glaciers,
overflows the land, in fact, down the valleys and fjords of Greenland,
by force of the superincumbent weight of snow, just as does the grain
on the floor of a barn (as admirably described by Mr. Jamieson)
when another sackful is emptied on the top of the mound already
on the floor. " The floor is flat, and therefore does not conduct the
grain in any direction; the outward motion is due to the pressure
of the particles of grain on one another; and, given a floor of infinite
extension, and a pile of sufficient amount, the mass would move
outward to any distance; and with a very slight pitch or slope it
would slide forward along the incline." To this let me add that if
the floor on the margin of the heap of grain was undulating, the
stream of grain would take the course of such undulations. The
want, therefore, of much slope in a country, and the absence of any

---

[1] While, for the sake of illustration, speaking of ice as " viscous matter," I must
not be understood as giving support to the " viscous theory " of glacier motion.

great mountain-range, are of very little moment " to the movement of land-ice, *provided we have snow enough*." [1]

As the ice reaches the coast it naturally takes the lowest level. Accordingly it there forks out into glaciers or ice-rivers, by which means the overflow of this great ice-lake is sent off to the sea. The length and breadth of these glaciers varies according to the breadth or length of the interspace between the islands down which it flows.[2] If the land projects a considerable way into the great ice-lake, then the glacier is a long one; if the contrary is the case, then it is hardly distinguished from the great interior ice-field, and, as in the case of the great glacier of Humboldt in Smith Sound, the interior ice may be said to discharge itself almost without a glacier. The face of Humboldt's glacier is in breadth about 60 miles. This, therefore, I take to be the interspace between the nearest elevated skirting land on either side. It thus appears that, between the inland ice and the glacier, the difference is one solely of degree, not of kind, though, for the sake of clearness of description, a nominal distinction has been drawn. The glacier, as I have said, will usually flow to the lowest elevation. Accordingly it may take a valley, and gradually advance until it reaches the sea. In the course of ages this valley will be grooved down until it deepens to the sea-level. The sea will then enter it, and the glacier-bed of former times will become one of those fjords which indent the coast of Greenland and other northern countries often for many miles; or these may be much more speedily produced by depression of the land, such as I shall show is at present going on. By force of the sea the glacier proper will then be limited to the land, and its old bed become a deep inlet of the sea, hollowed out and grooved by the icebergs which pass outwards, until in the course of time, by the action of a force which I shall presently describe, the fjords get filled up and choked again with icebergs, in all probability again to become the bed of some future glacier stream.[3] Where there is no fjord at hand, or where these defluents are not sufficient to draw off the surplus supply of ice, the " inland ice " will " boil " over the cliffs, overflowing its basin, and appear as hanging glaciers, whence every now and again huge masses of ice (the aërial equiva-

---

[1] 'Quart. Journ. Geol. Soc.,' xxiv. 1865, p. 166.

[2] Properly speaking, according to the ordinary nomenclature, the whole of the ice, from the " névé " downwards, should be called " glacier; " but as we have not yet penetrated sufficiently far into the interior to observe where the " névé " ends and the " glacier " begins, I have for the sake of distinctness adopted the above arbitrary nomenclature.

[3] The origin of fjords is more fully developed in Section iv. of this Memoir (p. 58).

lent of the bergs) are detached, as the attraction of gravity overcomes the cohesiveness of the ice. These have been seen and described by Dr. Kane on many parts of the Arctic coast. I noticed them in the shape of "miniature glaciers between the cliffs," ('Trans. Bot. Soc.' ix. 13) at Sakkak, lat. 70° 0′ 28″ N., and on the Waigat shore of Disco Island. In this latter locality they were the overflow of the inland ice *of the island*. They are also seen in the little local glaciers, where the bed they move in is shallow, and the seaward or outward end high, as near Omenak, where, however, I did not see them, but depend for my information on intelligent Danish officers resident in that section. In Alpine regions, away from the coast, the glacier, as it pushes its way down into warmer regions, either advances or retreats, according to the heat of the summer; but in either case it gives off no great masses of ice from its inferior extremity. The same is true of the Arctic glacier when it protrudes into some mossy valley without reaching the sea; but when it reaches the sea another force comes into operation. We have seen (1) the inland ice-field emptied by (2) the glacier; we now see the glacier relieving itself by means of (3) the iceberg or "ice mountain," as the word means.

3. *The Iceberg.*—When the glacier reaches the sea (fig. 1, e) it grooves its way along the bottom under the water for a considerable distance; indeed it might do so for a long way did not the buoyant action of the sea stop it. For instance, in one locality in South Greenland, in about 62° 32′ N. lat., between Fredrikshaab and Fiskernæsset, or a little north of the Eskimo fishing-station of Avigait, and south of another village called Tekkisok, is a remarkable instance of this. Here the "Iisblink," or the "ice glance" of the Danes (*i.e.*, the projecting glacier, though English seamen use the word iceblink in a totally different sense, meaning thereby the "loom" of ice at a distance), projects bodily out to sea for more than a mile. The bottom appears to be so shallow that the sea has no effect in raising it up; and the breadth of the glacier itself is so considerable as to form a stout breakwater to the force of the waves.[1] It was long supposed that the iceberg broke off from the glacier by the mere force of gravity: this is not so. It is forced off from the parent glacier by the buoyant action of the sea from beneath. The ice groans and creaks: then there is a crashing, then a roar like the discharge of a park of artillery; and with a monstrous regurgitation of waves. felt far from the scene of disturbance, the iceberg is launched into life. The breeze which blows out from the land,

---

[1] On this subject see also Nordenskjöld. *l. c.*, p. 364-5.

generally for several hours every day, seems, according to my observation, to have the effect of blowing the bergs out to sea; and then they may be seen sailing majestically along in long lines out of the ice-fjords. Often, however, isolated bergs or groups of bergs will float away south or north. Bergs from the ice-streams of Baffin Bay will be found in the southern reaches of Davis Straits; while others, bearing *débris* which could only have been accumulated in South Greenland, will be found frozen in the floes of Melville Bay, or Lancaster Sound. It is a common mistake, but one which a moment's reflection would surely dissipate, that bergs found in the south must all have come from the north, and that those further north must have come from the regions still farther northward. The winds and the currents waft them hither and thither, until by the force of the waves they break into fragments and become undistinguishable from the oozy fragments of floes around them. Often, however, they will ground either in the fjord or outside of it, and in this position remain for months, and even years, only to be removed by pieces calving or breaking off from them, and thus lightening them, or forced off the bank where they have touched bottom by the force of the displaced wave caused by the breaking off of a fresh berg. Ice much exposed to the sea only breaks off in small ice-calves, but not in bergs. This calving will sometimes set the sea in motion as much as 16 miles off. The colour of the berg is, of course, that of the glacier; but by the continuous beating of the waves on it the surface gets glistening. The colour of the mass is a dead white, like hard-pressed snow, which in reality it is, while scattered through it are lines of blue. These lines are also seen in the glacier on looking down into the crevasses, or at the glacier-face, and are in all probability caused by the annual melting and freezing of the surface-water of the glacier. Then another fall of snow comes in the winter; then the suns of summer melt the surface to some slight extent; this freezes, forming an ice different in colour from the compressed snow-ice of the glacier, and so on. I am aware, however, that this is a subject of controversy; and this view of mine is only brought forward as a probable explanation, suggested to me as far back as 1861, when I first saw glaciers in the upper reaches of Baffin Bay and on the western shores of Davis Strait, and long before I was aware that this streaked or veined character of glacier-ice had been a subject of dispute.[1]

---

[1] These blue stripes are several feet in dimension, and in them are generally found the "dirt bands" of foreign matter (stones, gravel, clay, &c.), the remains of the moraine. Dr. Rink thinks that the blue stripes are formed by a filling up of the fissures in the inland ice with water—"perhaps mixed with snow, gravel,

The greater portion of these bergs form long "streams" opposite their " ice-fjords," these streams being constantly reinforced by fresh additions from the land, poured out from the fjord. Hence certain localities in Greenland are distinguished by their " ice-streams ;" these localities being invariably opposite the mouths of ice-fjords. or fjords with great glaciers at their landward end pouring out icebergs. Few, if any, as I have already stated, are found on the east coast; but on the west (or Davis Strait and Baffin Bay side, from south to north, in the Danish possessions), the following localities, among others, chiefly known by their native names, are situated :—

1. Sermilik ice-fjord and ice-stream in about N. lat.	..	60	30
2. Sermeliarsuk	.. ..	61	32
3. Narsalik	.. ..	61	57
4. Godthaab	.. ..	64	30
5. Jakobshavn	.. ..	69	12
6. Tossukatek	.. ..	69	48
7. Great Kariak	.. ..	70	26
8. Little Kariak	.. ..	70	36
9. Sermelik	.. ..	70	41
10. Itifliarsuk	.. ..	70	52
11. Innerit	.. ..	70	56
12. Great Kangerdlursoak	.. ..	71	25
13. Upernivik	.. ..	72	57[1]

We have now sketched the ice-field with the glacier and the ice-berg. Are there no other defluents of the " inland ice?" This leads us to speak of :—

4. *The Subglacial Stream.*—What is under the inland ice is, I fear, a question we shall never be able to answer. No doubt the country is undulating; for I believe this immense glaciation overspread the country after the close of the Tertiary period, perhaps about the same period when Scotland lay under the ice cap. Continuously grinding over these rocks, a creamy mud must be formed, which mud must now be of considerable thickness, if not swept into hollows or washed out from beneath the ice. In the Alps the glacier is said to wear for itself a muddy bed, which Agassiz[2] calls *la couche de boue* or *la boue glaciaire,* and other authors *la moraine profonde*

and stones; and such a refrigeration of the water in the fissures may be supposed to be an important agency in setting in motion these great mountains
of ice."

[1] Rink: Om den geographiske Beskaffenhed af de danske Handels distrikter
i Nord-Grönland ; ndsigt over Nord-grönlands Geognosie. Det Kongl. danske
Vidensk. Selskab. Skr., 3 Bind, 1853, p. 71, *et lib. cit.* Dr. Rink altogether
resided for sixteen winters and twenty-two summers in Greenland.

[2] ' Études sur les Glaciers et Système Glaciaire,' p. 574.

(fig. 1, *b*); so that, I think, there can be little doubt that the Greenland inland ice has triturated down a similar clayey bed. However, another instrument in the arrangement, and, if I may use the term, "utilisation" of this mud, this *moraine profonde*, comes into play. Rink[1] has calculated the yearly amount of precipitation in Greenland in the form of snow and rain at 12 inches, and that of the outpour of ice by its glaciers at 2 inches. He considers that only a small part of the remaining 10 inches is disposed of by evaporation, and that the remainder must be carried to the sea in the form of sub-glacial rivers. These subglacial rivers are familiar in all Alpine countries, and in Greenland pour out from beneath the glacier, whether it lies at the sea or in a valley, and in summer and winter. He also mentions a lake adjacent to the outfall of a glacier into the sea, which has an irregularly intermittent rise and fall. "Whenever it rises, the glacier-river disappears; but when it sinks, the spring bursts out afresh,"—showing, as he thinks, a direct connection between the two. Arguing from what has been observed in the Alps, he concludes that an amount of glacier-water equivalent to 10 inches of precipitation on the whole surface of Greenland is not an extravagant hypothesis; and he accounts for its presence partly by the transmission of terrestrial heat to the lowest layer of ice, and partly by the fact that the summer heats are conveyed into the body of the glacier, while the winter cold never reaches it. The heat melts the surface-snow into water, which percolates the ice, while the cold penetrates a very inconsiderable portion of the glacier, whose thickness exceeds 2000 feet. As in the Alpine glaciers, these subglacial rivers are thickly loaded with mud from the grinding of the glacier on the infrajacent rocks; in fact, from the washings of the *moraine profonde*. This stream flows in a torrent the whole year round, and in every case which I know of (in the Arctic regions) reaches the sea eventually, though, no doubt, parting on the way with some small amount of its suspended mud. After it reaches the sea it discolours the water for miles, finally depositing on the bottom a thick coating of impalpable powder. When this falls in the open sea it may be scattered over a considerable space; but when (as in most cases) it falls in narrow long fjords, it collects at the bottom, shoaling up these inlets for several miles from their heads, until, in the course of time, the fjord gets wholly choked up, and the glacier seeks another outlet or gets choked up with bergs, which slowly plough their way through the deep banks of clay, until they get so consolidated together as to shut off the land alto-

---

[1] 'Naturhistorisk Tidsskrift,' 3rd series, vol. i. part 2 (1862), and 'Proc. Roy. Geog. Soc.,' vii. 76.

gether.[1] Supposing that the deposit only reaches 3 inches in the year, there is a bank or flat 25 feet thick formed in the course of a century. However, any one who has seen these muddy sub-glacier streams, and the way in which they deposit their mud, must be convinced that this estimate is far below the mark, and that an important geological deposit, which has never been rightly accounted for (if even noticed, as far as my observation goes), is forming off the coast of Greenland and wherever its great glaciers protrude into the deep quiet fjords. It ought also to be noticed that the fjords which have been the scenes of old ice-streams, in almost every instance end in a valley at the head, this valley being due, *first*, to the glacier which reclined on it and hollowed it out and, *secondly*, and further down, to the filling up of it by the glacier-clay. This form of fjord is not only common in Greenland, but also in every other part of the world where I have studied their form and formation.

After carefully examining and studying this clay, *I can find no appreciable difference between it and the brick-clay, or fossiliferous Boulder-clay*. Mr. Milne Home,[2] among other arguments against the theory that Boulder-clay has been formed by land-ice, remarks that he saw nothing forming in Switzerland at all comparable to Boulder-clay. Reserving to ourselves a doubt on that subject, I can only say that long after my opinion regarding the identical character of the subglacial-stream-clay and the fossiliferous brick-clay was formed, a very illustrious Scandinavian Arctic explorer visited Edinburgh and declared, as soon as he saw the sections of Boulder-clay exhibited near that city, that this was the very substance he saw forming in under the Spitzbergen ice. Many theoretical writers, however, confound the ordinary non-stratified azoic clay, and the finer, stratified fossiliferous clay.

In this clayey bed the Arctic Mollusca and other marine animals find a congenial home, and burrow into it in great numbers. However, as new deposits are thrown down, they keep near the surface, to be able to get their food; so that if to-day a catastrophe were to overwhelm the whole marine life of the Arctic regions, it would be found (supposing by upheaval or otherwise we were able to verify the fact) that the animals would only be imbedded in the upper strata of clay, and that the bottom one, with the exception of a few dead shells, would be azoic; yet I need not say how erroneously we should argue if, from this, we drew the inference that,

---

[1] I am glad to find that, independently, this identical view is held by Mr. J. W. Taylor, who resided for several years in Greenland, 'Proc. Roy. Geogr. Soc.,' v. p. 90 (1861).
[2] 'Trans. Roy. Soc. Edin.,' vol. xxv. p. 661; and ' Estuary of the Forth ' (1871).

at the time the bottom layers or strata of this laminated clay were
formed, there was no life in the Arctic waters, or that they were
formed under circumstances which prevented their being fossiliferous. The bearing of this on the subject in question need scarcely
be pointed out. It ought to be noted that, supposing we were able to
examine the bottom of the Arctic Sea (Davis Straits, for instance),
it would be found that this clayey deposit would not be found over
the whole surface of it, but only over patches. For instance, all of
the ice-fjords would be found full of it to the depth of many feet,
shoaling off at the seaward ends; and certain other places on the
coast would be also covered with it; but the middle and mouth of
Davis Straits and Baffin Bay, and the wide intervals between the
different ice-fjords, would either be bare or but slightly covered
with small patches from local glaciers; yet we should reason most
grievously in error, did we conclude therefrom that the other
portions of the bottom, covered with sand, gravel, or black mud,
were laid down at a different period from the other, or under other
different conditions than geographical position. These ice-rivers
seem, in the first place, to have taken their direction according to
the nature of the country over which the inland ice lies, and latterly
according to the course of the glaciers. No doubt they branch
over the whole country like a regular river-system.[1] When the
glacier reaches the sea, the stream flows out under the water, and,
owing to the smaller specific gravity of the fresh water, rises to the
surface, as Dr. Rink describes, "like springs"—though I do not
suppose that he considers (as some have supposed him to do) that
that water was in reality spring-water, or of the nature of springs.

---

[1] It may be somewhat superfluous for me to say that these subglacial streams
are totally different in nature from the streams which flowed in the old water-
courses found under the drift in various parts of the world. These were the beds
of the preglacial rivers, and are known to miners as "sand-dykes," "washouts,"
&c. On the North Pacific slope of the Rocky Mountains they are very common,
and are eagerly sought for by the gold-miners, the "old beds" generally yielding
a considerable amount of gold. In California, so thoroughly have they been
explored by the gold-diggers that, if proper records had been kept, a map of the
preglacial rivers might now be drawn, almost as detailed as that of the postglacial
or present river-system. The courses of these ancient rivers appear to have been
generally in the same direction, and to have had their outlets in the valleys near
about the same places as the present rivers. Sometimes these channels seem to
cross nearly at right angles. The old Yuba channel, for instance, when its course
was interrupted and diverted, ran through the site of the present village of
"Timbuctoo," crossing the bed of the present riverat Park's Bar; thence running
in a north-westerly course, and falling into the Rio de las Plumas (Feather
River), near Oroville, a considerable distance from its present junction with that
river at Marysville. These old channels exhibit the same windings and precipitous falls as the present river; and they have been cut in various places by
cañons and ravines; and portions of the older deposit, carried down, mingle with
the loose gravel and sand detached by more recent aqueous action.

Here are generally swarms of Entomostraca and other marine animals, which attract flights of gulls, which are ever noisily fighting for their food in the vicinity of such places.

We lived for the greater portion of a whole summer at Jakobshavn, a little Danish post, 69° 13′ N., close to which is the great Jakobshavn ice-fjord, which annually pours an immense quantity of icebergs into Disco Bay. In early times this inlet was quite open for boats; and Nunatak (a word meaning a " land surrounded by ice ") was once an Eskimo settlement. There is (or was in 1867) an old man (Manyus) living at Jakobshavn whose grandfather was born there. The Tessiusak, an inlet of Jakobshavn ice-fjord, could then be entered by boats. Now-a-days Jakobshavn ice-fjord is so choked up by bergs that it is impossible to go up in boats, and such a thing is never thought of. The Tessiusak must be reached by a laborious journey over land; and Nunatak is now only an island surrounded by the inland ice, at a distance—a place where no man lives, or has, in the memory of any one now living, reached. Both along its shore and that of the main fjord are numerous remains of dwellings long uninhabitable, owing to it being now impossible to gain access to them by sea. The inland ice is now encroaching on the land. At one time it seems to have covered many portions of the country now bare. In a few places glaciers have disappeared. I believe that this has been mainly owing to the inlet having got shoaled by the deposit of glacier-clay through the rivers already described. I have little doubt that—Graah's dictum[1] to the contrary, notwithstanding—a great inlet once stretched across Greenland not far from this place, as represented on the old maps, but that it has also now got choked up with consolidated bergs. In former times the natives used to describe pieces of timber drifting out of this inlet, and even tell of people coming across; and stories yet linger among them of the former occurrence of such proofs of the openness of the inlet.[2]

[1] ' Reise til Ostkysten af Grönland,' 1832, and translated by Macdougall, 1837.
[2] " There is another bay which I could not investigate to its bottom on account of the immense masses of ice that were setting out, and which is called by the natives Ikak and Ikarsek (Sound). It runs between Karsarsuk and Kingatok, and its length is from Karsarsuk to its end about 15 German miles; it is situated in 72° 48′, and the sea, at its entrance, is covered by numerous islands. All the natives living in this neighbourhood assured me unanimously that there had been a passage formerly to the other side of the land. They told me also that they were afraid that, with heavy north-easterly gales, the ice would go off again, and that the people from the other side, whom they describe as barbarians, would come over and kill them. They stated that, from time to time, carcasses of whales, which had been killed on the other side, pieces of wood, and fragments of utensils, were to be seen driving out of this bay."—Giesecke in Appendix to Scoresby's ' Journal of a Voyage to the Northern Whalefishery,' p. 168. Owing to an erroneous note and reference obtained at secondhand, I made it appear, in the

All that we know is, that such a transcontinental passage, if ever it existed, is now shut up. The glacier and the ice-stream have not changed their course, though, if the shoaling of the inlet [1] goes on (and if the glacier continues at its head, nothing is more certain), then it is just possible that the friction of the bottom of the inlet may overcome the force of the glacier, and that the ice may seek another course. As the neighbourhood is high and rocky, this is hardly possible with the present contour of the land. At the present day, the whole neighbourhood of the mouth of the glacier is full of bergs; and often we should be astonished on some quiet sunshiny day, without a breath of wind in the bay, to see the "ice shooting out" (as the local phrase is) from the ice-fjord, and to make up with the little bay in front of our door in Jakobshavn Kirke covered with huge icebergs, so that we had to put off our excursion to the other side of the inlet; and the natives would stand hungry on the shore, as nobody would dare put off in his kayak to kill seals, afraid of the falling of the bergs. In a few hours the bay would be clear, until another crop sprang out from the fjord. At any time it would be dangerous to venture near these bergs; and the poor Greenlander often loses his life in the attempt, as the bergs, even when aground, have always a slight motion which has the effect of stirring up the food on which the seals subsist. Accordingly the neighbourhood of these bergs is favourable for seals, in the attempt to capture which the hapless kayaker not unfrequently loses his life by falling ice. When we would row between two to avoid a few hundred yards' circuit, the rower would pull with muffled oars and bated breath. Orders would be given in whispers; and even were Sabine's gull or the great auk to swim past, I scarcely think that even the chance of gaining such a prize would tempt us to run the risk of firing, and thereby endangering our lives by the reverberations bringing down pieces of crumbling ice hanging overhead. A few strokes, and we are out of danger:

original paper of which this memoir is a partial reprint, as if this fjord spoken of in the preceding extract was Jakobshavn fjord, and that Jakobshavn fjord was open to boats in Giesecke's day. The error was of no great importance, but I have to thank Prof. Nordenskjöld for calling my attention to it. There is a tradition among the whalers that a whale was " struck " on the East Coast of Greenland, near Scoresby Sound, and was killed a few hours afterwards near Omenak Fjord, with the same harpoon in it—a certain proof of a passage across—*if the statement is true.* Perhaps, owing to the fewer icebergs on the East Coast, Franz Joseph's Fjord or Scoresby Sound may be the open easterly termination of one of these fjords now closed by ice on the west side. See, on the question of the former or present connection of the fjords on the East and West Coasts, Saabye's 'Greenland' (English Trans., 1818,) pp. 98-107.

[1] These inlets are, in fact, the " friths " of these ice-rivers Indeed, the term is actually used by some authors.

and then the pent-up feelings of our stolid fur-clad oarsmen find
vent in lusty huzzahs! Yet, when viewed out of danger, this noble
assemblage of ice palaces, hundreds in number being seen at such
times from the end of Jakobshavn Kirke, was a magnificent sight;
and the voyager might well indulge in some poetic frenzy at the view.
The noonday heat had melted their sides; and the rays of the red
evening sun glancing askance among them would conjure up fairy
visions of castles of silver and cathedrals of gold floating in a sea of
summer sunlight. Here was the Walhalla of the sturdy Vikings,
here the city of the sun-god Freyr, Alfheim, with its elfin caves,
and Glitner, with its walls of gold and roofs of silver, Gimle, more
brilliant than the sun, Gladsheim, the home of the happy, and there,
piercing the clouds, was Himlenberg, the celestial mount, where the
bridge of the gods touches heaven.[1] Suddenly there is a swaying, a
moving of the water, and our fairy palace falls in pieces, or with an
echo like a prolonged thunder-peal, it capsizes, sending the waves
in breakers up to our very feet. Some of these icebergs are of
enormous size. Hayes calculated that one stranded in Baffin Bay, in
water nearly half a mile in depth contained about 27,000,000,000 cu-
bical feet of ice, and must have weighed not less than 2,000,000,000
tons.

It is most probable that the cause of this "shooting out" of
bergs from the ice-fjord of Jakobshavn is due to the force gene-
rated by the detachment of a fresh berg from the glacier at the
extremity of the fjord. Occasionally, at the time of this "shooting
out," the waters of Jakobshavn harbour (a little fjord, the locality
of a now extinct glacier) will rise and fall with such tremendous
force as to snap a ship's cable. Actually the cable of the 'Mari-
anne,' a brig of 200 tons, was so broken in 1866. This wave is well-
known to the Greenland Danes, under the name of the 'kaäneël.'[2]
Various theories are afloat about it and its cause, which is
not very well known; but as it only happens when the ice is
"shooting out" in great quantities, it is most likely caused by the
displacement of the volume of water confined in the inlet; and
this wave is also felt outside; but its force is lost in the open
sea. It is also exhibited at Omenak and other harbours, when the
ice is shooting out of the ice-fjords in their vicinities; but these
harbours being situated at a greater distance from the scene of
action, it is not so much felt as at Jakobshavn, close to the ice-
fjord. From November to June, the fjords being frozen, there is
no "shooting out" of bergs but in July, and more especially in

[1] Hayes, op. c. p. 24.   [2] I spell the word phonetically.

August, and on until late in autumn, they pour out in great numbers. In concluding what I have got to say regarding the subglacial rivers, I cannot help remarking that the effect of this great ice-covering over Greenland must be to thoroughly denude any soft sedimentary strata which might have reclined on the underlying igneous rocks at the time when the whole country got so over-spread. Now we know that during the later Miocene epoch the country supported a luxuriant vegetation, as evinced by the remains which I and others have collected from these beds.[1] I was struck, when studying this subject in Greenland, with the fact (though I have no desire to push the theory too far) that the only places where I did not see former ice-action were the very localities where these Miocene beds repose. These localities are a very limited district on either side of the Waigat Strait, on Noursoak Penin-sula, and Disco Island, neither of these localities having apparently been overlain at any time by the great inland ice. Noursoak Penin-sula juts out from the land, and only nourishes small glaciers of its own ; and Disco Island is high land, possessing a miniature inland ice or *mer de glace*, with defluent glaciers of its own. If the great inland ice had ever ground over this tract, I hardly think it possible that the soft sandstone, shales, and coal-beds could have survived the effects of this ice-file for any length of time.

5. *The Moraines.*—Moraines are usually classified as *lateral, median, terminal*, and *profonde*,[2] or under the glacier. From the simple character of the Greenlander glacier, as described, it will be readily seen that the median moraine, formed by the junction of two lateral moraines, must be rare, while the terminal takes, ex-cept in rare instances, another form. Ordinary Alpine glaciers, when grinding down between the two sides of a mountain-gorge, get accumulated on their sides rubbish, such as earth, rocks, &c., which fall either by being undermined by the glacier, by frost, or by land-slips, until two lateral moraines are formed. If the glacier anastomoses with a second, it is evident that two of the lateral moraines will unite in the common glacier into a median one. When the glacier terminates, this moraine, carried along with it, is deposited at its base, and forms the terminal moraine. Over the

---

[1] Heer. in the ' Philosophical Transactions, 1869,' pp. 445-188. In this treatise of Prof. Heer I have printed a few notes on the geology of these Miocene beds ; but, owing to an accident, I did not see them in proof. Hence there are several errors. The title of the paper is also apt to mislead. These geological and other points I have since corrected in a full account of the geology of the Waigat Straits, &c., with illustrative map ('Trans. Geol. Soc., Glasgow,' vol. v. part i. p. 55).

[2] The term *moraine profonde* was first used by Hogard in his ' Coup d'œil sur le terrain erratique des Vosges' (1851), p. 10.

lower face of a glacier, according to the heat of the day, some material is always falling, a thimbleful of sand, it may be, trickling
down in the stream of water; or a mass of stone, gravel, and earth
may thunder over the edge. If the glacier advances, it pushes this
moraine in front of it, or, it is possible, may creep over it and carry
it on as a *moraine profonde.* This *moraine profonde* consists of the
boulders, gravel, &c., which the glacier, grinding along, has carried
with it, and which, adhering to its lower surface, help to grind
down infrajacent rocks, and at the same time get grooved in a corresponding direction. If the Greenland glacier does not reach the
sea, then the programme of the Alpine glacier is repeated; but
when the lower end breaks on reaching the head of the fjord, then
a different result ensues. The terminal moraine (*if there is any;*
for none comes over the inland ice, which leads me to believe that
it does not rise in mountains; and often the glacier is so short as
to take little or none from the sides of its valley) floats off on the
surface of the iceberg, and the *moraine profonde* either drops into
the sea, or is carried further on in the base of the iceberg: very
frequently this *moraine profonde* is composed of boulders and gravel,
and it is rare that they are not dropped before the berg gets out of
the fjord. The berg itself very often capsizes in the inlet, and deposits what load it may have on its surface or bottom at the bottom
of the sea; and when it gets out of the inlet, as I have already
described, it often ranges itself in the outside ice-stream; and if it
there capsizes, then the boulders lie on the bottom *there,* so that,
if the floor of the sea were raised up, a long line of boulders would
be found imbedded in a tenacious bed of laminated clay, with fossil
shells and remains of other Arctic animals, skeletons of seals, heaps
of gravel here and there, and so on, in what would then be a mossy
valley, most likely the bed of some river. Again, allow me to remark that a berg may not capsize by pieces breaking off from above
the water, but it may also lose its equilibrium (as is well known)
by being worn away, as is most frequently the case, at the base,
or (as is less known) by pieces calving off from below. If the berg
ground on a bank or shoal, or in any other water not deep enough
for its huge bulk to float in, it will often bring up from the bottom
boulders, gravels, &c., deposited by former bergs, and carry them on
until this material is deposited elsewhere; when grounding, it will
graze over the submerged boulders, or rocks just under water,
grooving them in long grooves; for an iceberg, it cannot be too often
remembered, is merely a mountain of ice floating in the sea. In my
earlier voyages in the Arctic regions I was rather inclined to under-
rate the transporting-power of bergs, as I saw but few of them with

any earth, rocks, or other land-matter on them. Though still believing that this has been exaggerated to support their theories by some writers, ignorant, unless by hearsay, of the nature of icebergs,[1] I am inclined to think that I was in error.

Towards the close of my voyage, in 1861, I had occasion to ascend to the summit of many bergs when the seamen were watering the vessels from the pools of water on their summits; and I almost invariably found moraine, which had sunk by the melting of the ice into the hollows, deep down out of sight of the voyager sailing past, but which would have been immediately deposited if the berg had been capsized. In 1867 I saw many bergs with masses of rock on them, and only at the mouth of Waigat one with a block of trap (?) so large, that it looked, even at a distance, like a good-sized house. The Greenland glaciers—or defluents of the inland ice—carry little moraine. The terminimal moraines are therefore little marked in comparison with what a glacier of the same size would deposit in the Alpine or other mountain regions abounding in glaciers. Indeed, the Swiss glaciers in almost no degree represent, even on a small scale, the great Greenland glaciation. It is unique.

6. *Life near the Ice-fjords.*—In the immediate vicinity of the Jacobshavn ice-fjord (and I take it as the type of the whole) animals living on the bottom were rare, except on the immediate shore or in deep water; for the bergs grazed the bottom in moderately deep water to such an extent as almost to destroy animal and vegetable life rooted to the bottom. In this vicinity bunches of algæ were floating about, uprooted by the grounding bergs; and the dredge brought up so little material for the zoologist's examination that, unless in deep water, his time was almost thrown away. Again, the heads of the inlets, unless very broad and open to the sea, are bare of marine life, the quantity of fresh water from the sub-glacial stream and the melting bergs being such as to make the neighbourhood (as in the Baltic) unfavourable for sea-animals. Some inlets are said to be so cold that fish leave them. I have not been able to confirm this in the Arctic regions. When stream-emptying lakes fall into the head of these fjords, having salmon in them, then seals ascend into the lakes in pursuit of them. Other localities, owing to the capricious distribution of life, would be barer

[1] I have found, however, that much of the "discoloration" in bergs is caused by the brown leaves of the *Cassiope tetragona* and other plants, growing among the rocks abutting on the glaciers, and blown down upon them. The supposed influence of icebergs in dispersing plants by carrying their roots and seeds in moraine I have shown to be in reality very little.—('Ocean Highways,' 1873.)

or more abundantly inhabited. Again, in shallow inlets, except for
Crustacea or other free-swimming animals, the bottom, continually
disturbed by the dropping of moraine or the ploughing up of bergs,
would be unfavourable for life. Accordingly, if the bed of the
Arctic Ocean in these places were raised, and we found the mouth
of a valley with laminated beds of clay rich in Arctic shells, and
the head bare of life, but still showing that the beds had been
assorted by marine action, supposing we were (as in Scotland)
ignorant, except by analogy, of the history of this, should we not
feel justified in saying that the beds at the one place and the other
were deposited under different conditions, and were in all likelihood of different ages? How just that apparently logical inference
would be I need scarcely ask.

5. ACTION OF SEA-ICE.

We have in the previous section in the most outline form sketched
the subject of Greenland glacial action. As the object of this paper
is not to form a summary of our knowledge on the subject, I have
not entered into a discussion of any points on the physics of ice,
further than was necessary to a right understanding of the subject
in hand. Suffice it to say that all sea-ice forms originally from the
" bay-ice " of the whaler, as the thin covering which first forms on
the surfaces of the quieter waters is called, and that this " bay-ice "
is almost entirely fresh, the effect of Arctic freezing temperature
being to precipitate the salt. Hence, when we talk of the temperature requisite to freeze salt water, it is merely equivalent to saying
that this temperature is requisite for the precipitation of the saline
constituents of the water. The water of the Arctic Sea is, according to Scoresby, of the specific gravity $1\cdot0263$.[1] At this specific
gravity it contains $5\frac{3}{4}$ oz. (avoird.) of salt to every gallon of 231
cubic inches, and freezes at $28\frac{1}{2}°$ Fahr. The specific gravity of this
ice is about $0\cdot873$. To enter upon this subject, of which the above
is only the summary of a long series of experiments, is foreign to
the object of this paper. From this bay-ice is formed the floe,
from the floe the pack-ice, and other forms familiar to Arctic navigators. In the summer the ice in Davis Strait on either side breaks
up sooner than that in the middle of the Strait, which remains for

---

[1] In an interesting series of experiments by Dr. Walker of the *Fox* Expedition,
it was shown that the bay-ice was never entirely free from salt. If sea water is
frozen its specific gravity is $1\cdot005$, showing salts, especially chloride of sodium or
common salts. Fresh water is often frozen on the surface of the salt.— (' Journ.
Roy. Dublin Soc., 1860,' vol. ii. pp. 371-380.)

a considerable time, forming the "middle ice" of the whalers. Still, however, a narrow belt remains attached to the shore during a considerable portion of the summer. This is called by the Danes in Greenland the "iis fod," and by the English navigators the "ice-foot." As the spring and summer-thaws proceed, land-slips occur, and earth, gravel, and avalanches of stones come thundering down on the ice-foot, there to remain until it breaks off from the coast, and floats out to sea with its raft-like load of land-débris. As the summer's long sunlight goes on, the ice, worn by the sea, parts with its load; and this may be shortly after its leaving the lands or it may float tolerably far south. The ice-foot, however, rarely carries its load as far south as the mouth of Davis Strait; and sea-ice is seldom seen far out of the Arctic regions, while, as we all know, bergs often float far out into the Atlantic. Often fields of ice will float along and, like icebergs, graze the surface of rocks only a wash at low tides; and therefore its action might be mistaken for that of icebergs or land-ice. In other cases I have known the ice-foot, laden with débris, to be driven up by the wind and high-tides on to low-lying islands, spits, and shores, piling them with the load thus carried from distant localities, so that blocks of trap from the shores of Disco or the Waigat might be drifted up on the beach at Cumberland Sound or on the gneissose shores of South Greenland.

It has even been found that in shallowish water the ice will freeze to the bottom of the sea; and in such situations the gravel, blocks, &c., there lying will freeze in and be carried out to sea, to be deposited in course of time in a manner similar to the superincumbent loads of the ice-foot, though more speedy. The same phenomenon holds good of the Baltic. In the Sound, the Great Belt, &c., the ground-ice often rises to the surface laden with sand, gravel, stones, and sea-weed. Sheets of ice, with included boulders, are driven up on the coasts during storms and "packed" to a height of 50 feet. How easily such sheets of ice, with included sand, gravel, or boulders, may furrow and streak rocks beneath may be imagined.[1] The patches of gravel on the pack-ice are owing, I think, to portions of the gravel-laden ice-foot having got among the ordinary materials of the pack; for I do not think that ice formed in deep water, unless when it passes over rocks, and therefore may take up fragments of stone or earth, has any geological significance.

---

[1] Forchhammer in 'Bull. de la Soc. Géol. de France,' 1847,' t. iv. pp. 1182-83; Lyell's 'Principles' (11th Ed.), vol. i., p. 383.

E

The conclusions which we are forced to draw from what I have
said regarding the depositing-power of glacier-streams, bergs, and
sea-ice must be :—1. That the bottom of Davis Strait must be composed of *various* materials; 2. That particular materials must pre-
dominate in particular localities; 3. That the bottom in the vicinity
of ice-fjords and in fjords must be chiefly composed of clay, with
boulders, gravel, and earth either scattered over it or in patches;
4. That the mouth and centre of Davis Strait and various banks,
such as Rifkol, must be chiefly composed of earth, gravel, boulders,
&c., with little or none of the glacier-clay; 5. That life must not
be uniformly distributed through this bottom; 6. That though
the lines of travelled blocks, boulders rubbed by grounding bergs,
ice, or by being brought out as part of the *moraine profonde*, will
be found scattered over every portion of the sea, still they will
chiefly be found in the lines of fjords and of the iceberg-stream;
6. That the clayey bottom of deep inlets will be little disturbed,
while that of shallow ones will be grooved and torn up by grounding bergs, &c.

RISE AND FALL OF THE GREENLAND COAST.

It may be asked—Have we any data for the conclusions in the
foregoing paragraphs, further than logical inferences from observed
facts justify us in drawing? Yes, we have; for there has been a
rise of the Greenland coast, laying bare the sea-bottom, as just now
there is a fall going on. This fact is not new; on the contrary, it
is notorious, but has been much misunderstood. We have the
Danes telling us on the most irrefragable evidence that the coast is
falling, while the Americans who wintered high up in Smith Sound,
saw there, and in all the country they visited to the north of
Wolstenholme Sound, raised sea-beaches and terraces, and accordingly say that it is rising in that direction, while, in truth, both of
them are right, but not in the exclusive sense they would have us
to imagine. There *has been* a rise; there *is* a fall going on. We
now supply the proofs.

1. *Rise.*—In Smith Sound both Kane's and Hayes's expeditions
observed a number of raised terraces 110 feet above high tide-mark,
the lowest being 32 feet. These were composed of small pebbles,
&c. Hence they concluded that the coast *was* rising. I think it
can be easily enough shown that this is only a portion of the old
rise of the Greenland coast. The interval between this locality and
the Danish possessions, commencing at 73° N. lat., has been so little
examined either by the geographer or the geologist that we can

say nothing about it ; but more to the south, and along the whole extent of the Danish colonies, this raised portion of the sea-bottom is seen. The hills are low and rounded, and everywhere scattered with perched blocks, boulders, &c., many of them brought from northern or southern localities. In other localities, in the hollows or along the sea-shore, we see several feet of the glacier-clay (the "brick-clay," in fact) full of Arctic shells such as are now living in the sea, Echinodermata, Crustacea, &c., while in other places, as might be expected from what I have said, the clay is bare of life. This clay corresponds identically in many places with some of the "brick-clays" of Scotland, though, as might be expected from the difference these clays partake of from the different rocks the trituration of which has given origin to them, they are in some places of different shades of colouring. In this glacier-clay (or shall I call it upper laminated Boulder-clay ?) all the shells found are of species still living in the neighbouring sea, with the exception of *Glycimeris siliqua*, and *Panopæa norvegica;* but as both of these are found in the Newfoundland Sea, we may expect them yet to be shown to be living in Davis Strait.[1] I have seen this "fossiliferous clay" up to the height of more than 500 feet above the sea, on the banks overlooking glaciers. At the Illartlek glacier, in 69° 27' N. lat., this glacier-clay, deposited on the bottom of the sea by some former glacier, now formed a moraine ; and on the surface of the ice I picked up several species of shells which had got washed out by the streams crossing over the glacier face. This Illartlek glacier does not reach the sea ; but supposing (as is doubtless the case elsewhere) that this clay had fallen on a glacier giving off icebergs, then the shells deposited in the old sea-bottom would be again carried out to sea, and a second time transferred to the bottom of Davis Strait ! I found this clay everywhere along the coast and in Leer Bay, south-west of Claushavn ; in knots of this clay are found impressions of the Angmaksætt (*Mallotus arcticus*, O. Fabr.), a fish still quite abundant in Davis Strait.[2] However, though this glacier-clay was found everywhere along the coast, yet it should be noticed that this was chiefly when glaciers had been in fjords, &c., and that often for long distances it would be sparingly found only in valleys or depressions.

Other evidences of the rise of the Greenland coast are furnished

---

[1] Mörch in Tillæg No. 7 til Rink's 'Grönland,' Bind 2, S. 143.

[2] " In general, I may say," remarks Agassiz, when speaking of the closeness with which Tertiary fishes agreed with recent ones, " that I have not yet found a single species which was perfectly identical with any marine existing fish, except the little species (*Mallotus*), which is found in nodules of clay, of unknown age, in Greenland." I am convinced that the age I have given is correct.

by ruins of houses being found high above the water, in places where no Greenlander would ever think of building them now. On Hunde (Dog) Island, in the district of Egedesminde, there are said to be two such houses, and two little lakes with marine shells naturalised in them, and remains of fish-bones, &c., on the shores. I only heard this when it was too late, so that to my regret I had to leave the country without paying a visit to this remarkable locality.

2. *Fall.*—This has been long known; but it is only within the last thirty years that special attention has been drawn to the subject, chiefly by Dr. Pingel,[1] who passed some time in Greenland. The facts are tolerably well known, how houses are found jammed in by ice in places where they never would have been built by the natives, as Proven, and so on. It may, however, be as well to recapitulate these proofs.

Between 1777 and 1779 Arctander noticed that in Igalliko Fjord (lat. 60° 43' N.) a small rocky island, "about a gun-shot from the shore," was entirely submerged at spring-tides; yet on it were the walls of a house (dating from the period of the old Icelandic colonists) 52 feet in length, 30 in breadth, 5 in thickness, and 6 high. Fifty years later the whole of it was so submerged that only the ruins rose above the water. The settlement of Julianeshaab was founded in 1776 in the same fjord; but the foundations of the old store-house, built on an island called "The Castle," are now dry only at very low water. Again, the remains of native houses are seen under water near the colony of Fredrikshaab (lat. 62° N). Near the great glacier which projects into the sea between Fredrikshaab and Fiskernæsset, in 62° 32' N., there is a group of islands called Fulluarlalik, on the shores of which are the ruins of dwellings which are now overflowed by the tide. In 1758 the Moravian *Unitas Fratrum* founded the mission establishment of Lichtenfels, about 2 miles from Fiskernæsset (lat. 63° 4'); but in thirty or forty years they were obliged once, "perhaps twice," to remove the frames or posts on which they rested their large *omiaks*, or "women's" (seal-skin) "boats." The posts may yet be seen beneath the water.

To the north-east of Godthaab (lat. 64° 10' 36" N., long. 51° 45' 5" W.[2]) on a point called Vildmansnæs (Savage Point) by Hans Egede, in 1721-36, several Greenland families lived. These dwellings are now desolate, being overflowed at high tide. At Nappersoak,

[1] 'Proc. Geol. Soc.,' vol. ii. p. 208.
[2] According to observations by the late Capt. v. Falbe, of the Royal Danish Navy, furnished to me by Capt. H. L. M. Holm, of the Hydrographic Department, Copenhagen.

45 miles north of Sukkertoppen (lat. 65° 25′ 23″ N., long. 52° 45′ 25″ W.), the ruins of old Greenland houses are also to be seen at low water.

In Disco Bay I had another curious instance brought under my attention by Hr. Neilsen, at the date of my visit, Colonibestyrer of Claushavn. The blubber-boiling house of that post was originally built on a little rocky islet, about one-eighth of a mile from the shore, called by the Danes " Speck-Huse-Oe," and by the Eskimo " Kruwelenwak," which just means the same thing, viz. " Blubber-house Island." For many years the island had been gradually sinking, until, in 1867, the year of our visit, Hr. Neilsen had been under the necessity of removing the house from it, as the island had been gradually subsiding until the floor of the house was flooded at high tide, though, it is needless to say, sufficiently far above high-water mark when originally built. On another island in its vicinity the whole of the Claushavn natives used to encamp in the summer, for the treble purpose of drying seals' flesh for winter use, of getting free from disturbance by the dogs, and of getting somewhat relieved from the plague of mosquitoes; but now the island is so circumscribed that the natives do not encamp there, the space above water not allowing of room for more than three or four skin tents. These facts are sufficient evidence that the coast of Greenland is falling at the present time; and I doubt not that if there were observers stationed in Smith Sound for a sufficiently long time, it would be found that the coast is also falling there, though hitherto only Kane and Hayes have stayed there, but for too short a period to decide on the matter; and I cannot see that there is the slightest reason why the fall should halt at Kingatok (N. lat. 73° 45′), the most northern Danish post, and the most northerly abode of civilised man. Circumstances have only allowed of its being noted so far.

Hr. Neilson told me that he considered that Disco Island, opposite Claushavn, was rising, because the glaciers were on the increase. I think that if there is no more evidence than this for that supposed fact, we may lay it aside as erroneous, because the glaciers are undoubtedly increasing by the increase of the interior *mer de glace* on the island, and by the regular descent which they are making to the sea. Disco Island is a miniature edition of Greenland; it has its inland ice, its defluent glaciers, and its sub-glacial rivers, which sweep the denuded material from beneath the ice.

I have made an attempt to estimate the rate of fall; and though we have no certain data, yet I believe that it does not exceed 5 feet in a century, if so much; so that none of us will live to see Greenland overspread by the sea. Such at least are the views I

have arrived at from a careful study of this question. Little doubt
remains in my mind as to its correctness. The only serious reason
for hesitating to ask the reader to accept this elucidation of the
subject is, that it would appear that for some indefinite period
there has been a gradual elevation of most of the circumpolar
region going on. The facts in regard to this have been carefully
collated by Mr. H. Howorth,[1] though it must be acknowledged
with apparently a foregone conclusion, or at least a strong bias to
the doctrine he has espoused, and to his memoir the reader can be
safely recommended. One fact I may mention, which I am not
aware has been noticed by Mr. Howorth. A few years ago the
Norwegian walrus hunter discovered a group of small islets north
of Novai Semläi. They were merely sandy patches scattered with
boulders dropped from icebergs which had at one time floated over
them, raised but a few feet above the sea—

" . . . . islands salt and bare.
The haunt of seals and orcs and seamews' clang."

On some of the islets—notably on Hellwald's and Brown's—were
found West Indian fruits washed up by the Gulf Stream; hence
they were named "The Gulf Stream Islands." Yet only about
two centuries ago the Dutch took soundings on the very spot where
these islands have since been gradually raised above the sea. It is
also said that the whale (*Balæna mysticetus*) has left the Spitzbergen
Sea, owing to the waters having got too shallow for it, on account
of the gradual rise of the bottom. On Franz Joseph's Land there
are also raised beaches. The whole question is an important and
interesting one for the naturalists of the present Arctic Expedition
to attempt the solution of. Here I may point out what seems to be
a fallacy in the reasoning of those authors who write about the
denuding powers of rivers, and calculate that such and such a
country will be overwhelmed by the sea in so many millions of
years. Whatever the land loses by denudation the sea gains; and
therefore the two forces keep pace with each other. We thus see
in Greenland two appearances : (1) In the interior what Scotland
once was ; (2) on the coast what Scotland now is.

7. APPLICATION OF THE FACTS REGARDING ARCTIC ICE-ACTION AS EX-
PLANATORY OF GLACIATION AND OTHER ICE-REMAINS IN BRITAIN.

In the paper referred to,[2] and in the geological portion of the

---

[1] ' Journ. of the Roy. Geog. Soc.,' vol. xliii. (1873). p. 240.
[2] ' Quart. Journ. Geol. Soc.,' vol. xxvii. p. 671 ; also ' Popular Science Review,'
August, 1871, and April, 1875 ; and more popularly in Kingsley's ' Town
Geology,' pp. 48-52.

Royal Society's 'Manual of the Natural History of Greenland,' as well as in the instructions by the distinguished head of the Geological Survey of Great Britain—than whom there is no higher authority on the subject in Britain—will doubtless enter fully into the application of the foregoing facts as affording some explanation of the puzzling deposits of late geological age in Britain, and other portions of the northern hemisphere, and known as the "glacial beds" or remains. We are still far from understanding fully all the phenomena presented by these glacial remains.   Still, as it is only by the study of a country like Greenland, which is in a condition similar to that which Scotland and a great portion of the northern hemisphere are believed to have been during the glacial period, it may be well, though this is not the place for geological details, to briefly recapitulate the general conclusions which I have arrived at from the study of Greenland ice :—

(1.) The brick clays or laminated fossiliferous clays of Scotland, &c., are exactly the same as the clays now filling up the Greenland fjords from the mud-laden streams which flow from under the glaciers, and are due to the same or similar agents acting during the " Glacial period." These agents *must* have been acting at that period, and the clay formed from these sub-glacial streams has never yet been accounted for.

(2.) The non-fossiliferous " till," though there are still appearances in this non-stratified deposit that we cannot account for, is in all likelihood the representation of the *moraine profonde* of the great ice-cap. Had it been moraine dropped from icebergs, as has been argued, even supposing that icebergs could deposit it so uniformly over great tracts and to such a thickness, it would have been fossiliferous and stratified. It is neither. (3.) Kaimes, Osars,[1] Escars, &c., are only the "banks" of the old glacial seas. Some may be of fresh-water origin, but most are marine. (4.) The angular " travelled blocks " (the "foundlings" of the Swiss mountaineers) have been dropped by icebergs floating over the submerged country. The rounded ice-borne boulders are part of the *moraine profonde.*

The conclusions thus briefly summarised, with the deductions as to the former state of Scotland, will be found fully stated in the memoirs and works referred to. Lastly, the observer ought to guard against supposing that, in the old glacial seas or on the glacial lands, life was poor. If we are to judge the past by the present, we have no right to suppose any such thing.

The rarity of life in many of the glacial beds need not be

---

[1] A Swedish word so pronounced, but written *Asar* or *Aasar.*

wondered at when we consider the capricious and even sporadic distribution of life in the fjords of Greenland. It is possible also, as Lyell suggests, that animal life was originally scarce ; for " we read of the waters being so chilled and freshened by the melting of icebergs in some Norwegian and Icelandic fjords that the fish are driven away and all the mollusca killed."[1] He also points out most justly that, as the moraines are at the first devoid of life, if transported by icebergs to a distance, and deposited where the ice melts, they may continue as barren of every indication of life as they were where they originated. That the freshening of the water of fjords does destroy or prevent animal life developing, I have already shown ; but I doubt whether the chilling has much, if any, effect ; and the recent researches of Carpenter, Jeffreys, Thomson, and others, show that the idea which was suggested, that the sea might then be too deep for animal life, is without foundation : for life seems, as far as our present knowledge goes, to have no zero ; besides, the shells found in the glacial formations are not deep-sea shells. Again, we must be careful to avoid concluding that the plant- and animal-life on the dreary shores or mountain-tops of the old glacial Scotland was poor. In Greenland, the outskirting islands support a luxuriant phanerogamic vegetation of between 300 and 400 species of plants ;[2] the sea is full of fishes and invertebrates, which shelter in forests of Algæ. Plants even ascend to the height of 4000 feet. Millions of seals and whales, and of many species, sport in these waters, or are killed in thousands every spring on the pack-ice or land-floes. Every rock is swarming and noisy with the cries of water-fowl ; reindeer browse in countless herds in some of the valleys ; the Arctic fox barks its *hue! hue!* from the dreariest rocks in the depth of winter ; and the polar bear is on the range all the year round. Land-birds from southern regions come here for a nesting-place,[3] and from the snowy valleys the Greenlanders will bring in the depth of winter sledge-loads of ptarmigan into the Danish posts. Life is so abundant that the Danish Government find it profitable to keep up trading-posts there, and the collecting and preserving of the skins, oil, and ivory of the native animals afford profitable employment to a considerable population. Independently of the fish eaten, the seals

---

[1] Lyell's ' Antiquity of Man,' p. 268.

[2] The present writer, in little more than two months, amid many other occupations, collected on the shores and in the vicinity of Disco Bay alone, 129 species of flowering plants and vascular cryptogams, more than 40 mosses, 11 Hepaticæ, more than 100 Lichens, including many new species, about 50 Algæ, and several Fungi (see ' Transactions of the Edinburgh Botanical Society,' vol. ix.).

[3] About 115 species of birds are found in Greenland.

used as food and clothing, and the oil consumed in the country, it may not be irrelevant in this light to present the following list of a portion of the annual exports of the Danish settlements in 1855 :[1]—

9569 barrels of seal-oil.

47,809 seal-skins.

6346 reindeer-skins. There is on record the fact of 30,000 being exported in one year.

1714 fox-skins.

34 bear-skins (the animal being almost extinct in Danish Greenland).

194 dog-skins (in addition to the numerous teams used by the natives).

3437 lbs. rough eider-down.

5206 lbs. of feathers.

439 lbs. of narwhal ivory (the natives also using up much for their implements).

51 lbs. of walrus ivory (the walrus being little pursued).

And 3596 lbs. of whalebone (very few of the *Balæna mysticetus* being killed).

Add to this that, when the Danes came to Greenland first, there was a population not much less than 30,000; and to this day there lives within the Danish possessions a healthy, hearty race of upwards of 10,000 civilised intelligent hunters of narwhal, seal, and reindeer, with schools and churches within sight of the eternal inland ice, and with a long night of four months, which, perhaps, Scotland had not during the glacial epoch. I do not believe, however, that our shores were inhabited then; but still I see no reason why they *could* not have been ; and, with the bright skies and warm sunshiny days of a Greenland summer fresh in my memory, I cannot bring myself to believe in the poetically gloomy pictures pseudo-scientific writers have delighted to draw of the leaden skies, the misty air, and unutterable dreariness of our Scottish shores in that incalculably distant period when glaciers ran through our valleys from the inland ice, and icebergs crashed in our romantic glens, then fjords of that glacial coast.

<hr>

[1] For this return I am indebted to my friend Dr. Rink, the most eminent authority on all matters connected with Greenland. See also my monographs of Greenland Mammals in the ' Proceedings of the Zoological Society of London ' for 1868, and in ' Petermann's Geographische Mittheilungen,' 1869.

8. On the Formation of Fjords.[1]

Intersecting the sea-coasts of various portions of the world, more particularly in northern latitudes, are deep, narrow inlets of the sea, surrounded generally by high precipitous cliffs, and varying in length from 2 or 3 miles to 100 or more, variously known as "inlets," "canals," "fjords," and even, on the western shores of Scotland, as "lochs." The nature of these inlets is everywhere identical, even though existing in widely-distant parts of the world, so much so as to suggest a common origin. On the extreme north-west coast of America they intersect the sea-line of British Columbia to a depth, in some cases of upwards of 100 miles, the soundings in them showing a great depth of water, high precipitous walls on either side, and generally with a valley towards the head. On the eastern shore of the opposite Island of Vancouver no such inlets are found, but on the western coast of the same island they are again found in perfection; shewing that, in all probability, Vancouver Island was isolated from the mainland by some throe of Nature prior to the formation of the present "canals" on the British Columbia shore, but that the present inlets on the western shore of Vancouver Island formed, at a former period, the sea-board termination of the mainland, and were dug out under conditions identical with those which subsequently formed the fjords now intersecting the coast.

Jervis Inlet may be taken as the type of nearly all of these inlets here, as well as in other portions of the world. It extends in a northerly direction for more than 40 miles, while its width rarely exceeds 1½ mile, and in some places is even less. It is hemmed in on all sides by mountains of the most rugged and stupendous character, rising from its almost perpendicular shores to a height of from 5000 and 6000 feet. The hardy pine, where no other tree can find soil to sustain life, holds but a feeble and uncertain tenure here ; and it is not uncommon to see whole mountain sides denuded by the blasts of winter or the still more certain destruction of the avalanche which accompanies the thaw of summer. Strikingly grand and magnificent, there is a solemnity in the silence and utter desolation which prevails here during the months of winter, not a native, not a living thing to disturb the solitude ; and though in the summer a few miserable Indians may occasionally be met with, and the reverberating echoes of a hundred cataracts disturb the

---

[1] Abridged, with additions and corrections, from the 'Journal of the Royal Geographical Society,' 1869 and 1871.

silence, yet the desolation remains, and seems inseparable from a scene Nature never intended as the abode of man. The depths below almost rival the heights of the mountain summit: bottom is rarely reached under 200 fathoms, even close to the shore.[1] The deep inlets on the Norwegian coast, known as *fjords*—a familiar name, now applied generally to such breaks in the coast-line—are two well known to require description. On the coast of Greenland are again found similar Sounds, indenting both sides of that group of islands (?), but more particularly the western or Davis Strait shore. Most of these inlets are thickly studded with floating ice-bergs, and others are so densely choked with them as to receive the name of ice-fjords. All of these fjords form the highways by which the icebergs float out from the glaciers at their heads, when-ever these prolongations of the great *mer de glace* of Greenland (the " inland iis ") reach the sea. After a long and careful study of these fjords in most parts of the world where they are found, 1 have come to the conclusion that we must look upon glaciers as the material which hollowed them in such an uniform manner. Everywhere you see marks on the sides of the British Columbian fjords of ice-action;[2] and there seems no reason to doubt but that they were at one time the beds of ancient glaciers, which, grinding their outward course to the sea, scooped out these inlets of this great and uniform depth. At the time when these inlets formed the beds of glaciers, the coast was higher than now. We know that the coast of Greenland is now falling; and, supposing that the present rate of depression goes on, many glacier valleys will in course of time become ice-fjords. After having seen not a little of the abrading action of ice during three different visits to the Arctic regions, extending in circuit from the Spitzbergen Sea to the upper reaches of Baffin Bay and westward and southward to the " Meta Incognita " of Frobisher, I cannot side with those geologists who, judging ice-action merely from what is seen of the comparatively puny glaciers of the Alps and other European ranges, are inclined to under-estimate the abrading power of the glacier. 1 do not, however, for a moment pretend to assert that the valleys in which glaciers in the Arctic regions (or elsewhere) now lie were originally formed by the glacier. On the contrary, I am at one with those who believe that these rents were chiefly due to the

[1] 'Vancouver Island Pilot,' p. 139 (Admiral Richards).
[2] A fact which my friend, Dr. Comrie, R.N., whose familiarity with the British Columbian coast is well known, informs me that he has repeatedly confirmed. I am authorised to say that in his mind no doubt remains that these fjords were formed in the manner I have described.

volcanic disturbances which threw up the mountain ranges, and
that the glacier merely took advantage of the depression. However, by long abrasion it hollowed out the valley into the form we
now see it in the fjords under description. At this present day,
not far from the head of most of these inlets, glaciers are found in
the Coast Range and Cascade mountains in British Columbia; and
along both ranges marks of old glacier action can be seen 2000 to
3000 feet below their summits, and even near the sea-margin.
Such a depression of the coast, with the presence of the lower
temperature then prevailing, would fill those fjords with glaciers.

Such is the thesis I ventured to put forth on the nature of these
fjords or inlets. That it would be allowed to pass unchallenged
was scarcely to be expected, when such a variety of views were
held on the subject. As the object of these pages is not to pro-
mulgate the author's own views, but to give an unbiassed statement
of the doctrines held in regard to the subjects of them, I can
perhaps best serve the purpose I had in view, by simply giving the
reply to my various critics. By perusing this the reader can at
once see the arguments *pro* and *con.* the subject, and form his
own opinion as to which explanation most fully meets the difficulty,
and from this stand-point endeavour to aid in the solution of the
question.

The doctrines broached have been favourably received on the
Continent and in America, and by many of those in this country
best able to judge regarding their reasonableness. My paper, however, in so far as regards the theory of the formation of fjords, has
been honoured by two special attacks having been directed against
it. The first[1] of these in time is by Mr. Joseph W. Tayler, so long
connected with the cryolite mines of Arksut Fjord, in Greenland ;
the second[2] is by the late illustrious President of this Society.
Though no words coming from Sir Roderick Murchison on a subject
of physical geology can fail to be received with the careful attention and profound respect which his long and pre-eminent services
to science entitle them to, and though well aware of Mr. Tayler's
long residence in Greenland, yet, with every respect for both, I
must humbly submit that they have not made good their case for
the doctrine that glaciers have nothing whatever to do with the
formation of fjords. On the contrary, after having studied the
subject anew, and visited, since my paper was published, several of
the regions where fjords abound, and which are cited in illustration of my ideas in the paper mentioned, I am convinced—even

[1] 'Proceedings R. G. S.,' vol. xiv p. 156; 'Journal,' vol. xl. p. 228.
[2] Ibid. p. 327 ; 'Journal,' vol. xl. p. clxxiv.

more than before—that the explanation I then gave, if not exactly
the true one, is at least nearer the truth than the one opposed to it.
It is with a view to recapitulate these arguments, and not with a
view to bolster up a theory, which must eventually stand or fall on
its own merits, that I ask a place in the Society's transactions for
these additional remarks. The question is not so much whether
fjords were hollowed out by glaciers, but simply a renewal of the
contest between the rival schools of "catastrophists," who believe
that all the great physical features of the world have been caused
by some cataclysm or cataclysms of Nature; and of "uniformi-
tarians," who teach that the uniform and long-continued action of
the forces at present acting on the earth's surface would be sufficient to account for many features hitherto ascribed by their rivals
to huge throes of Nature. The whole subject has been discussed
over and over again, and all the main arguments which have been
brought to bear against this particular application of the uniformi-
tarian doctrines, have been advanced against some other application
of it, in explanation of other physical features. Nor have the
supporters of the contrary view been backward in replying; and
the whole matter stands *in statu quo*, or as the leanings of physical
geologists bear to one side or other of the controversy. Foremost
and chief of the school of catastrophists was our distinguished
President, and our Transactions almost yearly bear witness to the
skill, eloquence, and learning with which he has employed the
weapons of his party against the adherents of the opposite view.
. Originally, when he visited the Arctic Regions for the first time
ten years ago, a disciple of Sir Roderick in this country, and of
von Buch in Germany, the present writer must confess that additional observation and more extensive travel have led him to desert
to the enemy. The paper mentioned is a result of his studies
under the new banner, and these further remarks must be taken as
his justification of the faith that is in him. The arguments brought
against him both by Sir Roderick and Mr. Taylor are so nearly
identical, so far as they go, that he may be permitted to reply to
them conjointly. Had, however, Mr. Taylor waited until the
publication of my complete paper, he would have spared himself
and the Society some of his remarks, which his impatience for what
seemed an easy victory has induced him to advance against the imperfect statement of my case in the fragmentary report published
in the 'Proceedings.' Unfortunately he commences his arguments
by entirely misunderstanding my views.

1. *Glaciers and Fjords.*—When Mr. Taylor says that he "takes
it for granted" that by "hollowing" I mean causing fjords to be

where none were before, he takes for granted what I never did
grant. On the contrary, I have always shunned the extreme views
of either geological school, which would assign the origin of all
physical features *alone* to the causes of which they are the advocate.
I believe, and consider that in my paper I made it clear, that as a
glacier outpour in approaching the coast, or in falling from an
elevation, always takes the line of least resistance; so in former
times it sought the valleys and depressions then existing in the
coast-line of Greenland. It might even have taken the "gulches"
and ravines which former volcanic force had formed. But at that
time Greenland, Norway, and other fjord-indented countries did
not present the aspect they do now. Fjords, as we understand
them now, did not then exist. It was to the long-continued action
of the glaciers moving over these valley-beds that the deep uniform
inlets are due. Probably the sea assisted the glacier after the
coast had fallen, but that the sea alone cut out these fjords no one
who knows anything of the action of the waves on a coast-line can
for a moment entertain. If the rocks along a coast were alternately
soft and hard in parallel lines, then the sea by wearing away the
soft and leaving the hard, might accomplish the feat of forming
fjords. But as no coast is formed on this plan, then it must follow
that either the shore is equally worn away, according to the force
of the waves, or cut here and there into bays, of the rocks out of it.
At that time the present coast-line of these continents did not
exist, and when Mr. Tayler attempts to disprove my theory by
talking of the present fjords of Greenland as if they were of
primeval origin, I fear that he does not clearly understand the
doctrines held by all geologists, that the Greenland coast has been
undergoing a continual oscillation. He mixes up, with a curious
confusion of ideas, the fjords after they are formed and the causes
which formed them. A moment's consideration would convince
any one that the coast of Greenland at that time was entirely
different from now, and that since these fjords have been formed it
has undergone many changes of level.

2. *Grinding Power of Glaciers.*—When Mr. Tayler and Sir Roderick
Murchison inform us that ice has no abrading power, and only slides
over the rock, they will scarcely expect me to agree with them.
This question is as yet *sub judice*, though I am inclined to believe
that those who assert the grinding power of ice have made out a
very clear case. I cannot understand how any one who has seen
the rounded ice-planed hills of Greenland, and the immense mud-
laden stream which flows out from under every large glacier, as the
result of the grinding action of the ice, by means of its file-like

moraine profonde, can believe that the glacier merely slides over the surface of the rock without causing any abrading action. Though I cannot allow that Mr. Tayler's long residence in the Arctic regions —principally, I presume, in the vicinity of Arksut Fjord—enables him to come so positively to the conclusion he does, which is only the old theoretical opinion of some geologists, derived from the comparatively puny glaciers of the Alps; yet even there he must have seen the stream laden with clay, pouring from under every glacier, and choking up the neighbouring fjord, and shoaling even the open sea around. Where can this mud have come from, if not from the country underlying the glacier, and the great inland ice of the interior of Greenland; and if from these—as undoubtedly it has—how can any one, with these well-known facts before his eyes, declare that the glacier has little or no abrading power? Mr. Tayler, even when wishing to prove the contrary, states a fact which entirely cuts the ground from beneath his feet. "It is true," he says, "that boulders and *débris*, borne along by the ice, scratch, polish, and grind the rocks to a considerable extent; but, though strong as a transporting agency, ice alone has but little excavating power; it is like the soft wheel of the lapidary—the hard matter it carries with it does the polishing." Exactly so. It is to the geologist a matter of the most supreme indifference whether it is the ice of the glacier itself, or the *moraine profonde*, invariably accompanying it, which does the abrasion of the underlying rocks, *so long as it is done*. And that even Mr. Tayler, in contradiction of his own doctrine, seems to allow. Some opponents of the doctrine of ice-abrasion, who even allow less power to the glacier than Mr. Tayler, always lose sight of the long period during which the glacier must have been acting to form these long fjords or inlets of the sea as they now exist. This allowed—and there is no geologist who will doubt that though the glacial period is but of yesterday in geological time, viewed in the light of human chronology, it is so incalculably distant that it would be vain to attempt to calculate the date of that epoch, and even allowing that the glacier pouring down the Norwegian, British Columbian, or Greenland valley of that date, only removed every year by means of the sub-glacier stream, *one* inch of rock or other subcumbent stratum—it requires but a very moderate number of years to excavate the broadest and deepest fjord in the world. At that time the coast was higher than now, and it is the lowering of the coast, combined with the deepening of the valley, that has converted what was once a glacier-valley into a fjord or inlet of the sea. The great error of the catastrophists is, that their method of thought has led them unconsciously to expect the slow and uniform action of the forces of nature—acting from the beginning until now under the control of one uniform unchanged and unchangeable law—to act as rapidly as their "cataclysms" and other prodigious "catastrophes of nature." Mr. Taylor is especially, I think, a little unreasonable in disowning the abrading power of ice, because, in the eighteen years or so during which he was a witness of its power, he did not see it "hollow out" a fjord, and complete it ready for use! After all, the Roman poet, who eighteen hundred years ago saw the rain-drops splashing on the pavement of Tomi, had a clearer idea of the effect of the slow, but constant, action of the forces of nature than some geologists in later times—"Gutta cavat lapidem, non vi, sed sæpe cadendo." Therein lies the whole theory of ice and river action. When we see a smoothly-gliding river excavate cañons thousands of feet in depth through the solid rock, surely it would be inconsistent to deny that an ice-river flowing over the same spot for thousands of years, may, assisted by a huge file, in the shape of the *moraine profonde*, which it carries along with it on its under-surface, and the sub-glacial river to carry off part of the *débris* thus worn, do something approaching to this?

If the advocates of the non-abrading power of ice will not allow that glaciers can convert a valley in course of time into a deep glen, and that then, by the aid of an oscillation of the coast, the sea enters and the glacier floats away in icebergs, and its former bed now becomes the fjord through which they sail, I cannot expect them to give in their adhesion to Professor Ramsay's views regarding the excavation of lake-basins by means of glaciers.[1] On the contrary, this view of that distinguished geologist, while gaining many converts, has been violently attacked both in this country and on the Continent, yet, I venture to think, without being at all shaken in its main points. Already it has been extensively adopted; and only recently an eminent American naturalist—Professor Newberry, of New York—has applied it to account for the formation of the great American lakes.[2] Yet Professor Ramsay's theory requires much more of ice than is required of it by mine.

3. *Filling up of Fjords.*—When Mr. Taylor says that, "instead of glaciers excavating fjords, they are continually filling them up," he must not expect me to follow him; for here, again, he loses the thread of his arguments with a confusion of ideas which renders it

[1] 'Quarterly Journal, Geological Society,' vol. xviii. p. 185 (1862.
[2] 'Annals of the New York Lyceum of Natural History' (1869). Prof. Nordenskjöld, while agreeing that glaciers exercise an abrading influence, does not, however, coincide with Prof. Ramsay's theory of the formation of lake-basins (Nordenskjöld, *l. c.*, p. 365).

quite unnecessary to say more in refutation of what has nothing
whatever to do with the subject in hand. It does not at all follow
that, because *ancient* glaciers hollowed out the present fjords, the
sub-glacial stream flowing into them from *modern* glaciers may not
shoal them up. But the modern glacier, like the ancient one, whether
ending in the head of a fjord (the bed of an ancient glacier) or at
the open sea—as at the great "Iisblink," 15 miles north of Frederikshaab—is, I believe, unquestionably excavating out the valley in
which it lies, to become hereafter, in some future period of Greenland
history—either through a change of climate or of coast-level—a deep
valley or a deeper fjord. I know—as does any one at all acquainted
with Greenland—that this great glacier, though it is not the only
one, reaches the sea without entering a fjord, and finding the sea
too shallow to buoy its seaward end up, and so break it off in the
form of icebergs (as in the deep fjords), it pushes its way along the
bottom for some distance, until getting into deeper water it will
again, like the others, discharge its icebergs. This, again, is quite
foreign to the subject of the formation of fjords. There are *modern*
glaciers. I spoke of *ancient* ones. Still even the great "Iisblink"
spoken of, though it happens—accidentally it may be said—not to
enter a fjord, is, nevertheless, by the part of it which lies on land,
grinding down the infra-jacent country and acting the part of the
ancient glaciers which formed the *present* fjords. In regard to this
filling up of the fjords by the modern glaciers at their head, this is
due to the mud brought down by the sub-glacial stream, and which
is again due to the abrasion of the rocks by the incumbent glacier
moving over them. In another memoir, 'On the Physics of Arctic
Ice as explanatory of the Glacial Remains of Scotland,'[1] I have entered
into a full discussion of this and other points connected with Arctic
glaciers, so that it would be needless to take up space here with any
résumé of my observations. In that memoir I have estimated that,
at the very lowest calculation, this glacial mud is accumulating at
the *head* of these fjords at the rate of not less than 25 feet thick in a
century. Accordingly it has closed some old fjords with ice, the
glacier at their head being no longer able to discharge its bergs,
owing to the shallowness of the water, and in some cases, as previously pointed out by Dr. Rink[2] and Mr. Tayler,[3] the glaciers are
seeking new outlets, on the principle of ice seeking the plane of least

---

[1] ' Quarterly Journal of the Geological Society,' vol. xxiv. (1871) pp. 671-701 ;
and in a more condensed form, reprinted at pp. 27-58 of this ' Manual,' and " Das
Innere von Grönland," in Petermann's ' Geographische Mittheilungen,' October,
1871.
[2] ' Grönland Geographisk og statistisk beskrevet,' &c.
[3] ' Proceedings of the Royal Geographical Society,' vol. v. p. 93 (1861).

F

resistance. Mr. Tayler asks triumphantly, Why does not the great glacier referred to cut its way through the sand and *débris* which lie at its base? I answer, Give it time, and most assuredly it will do so. That is, however, not a question at all connected with the *abrading* power of the glacier. It is simply connected with the question of what mechanical force the glacier exerts in pushing forward. When the average rate of the downward and outward progress of a Greenland glacier is only about *five inches* per diem, we must not be in a great hurry to see the solution of the question Mr. Tayler has proposed. But, just as truly as a glacier moves, will this rubbish be shot into deeper water, and the end of the glacier, buoyed off by the deeper water, break off in the form of an iceberg.

4. *The Walls of Fjords.*—I am asked, Why were not the soft sand-stone, coal, "black-lead," &c., of which the sides of many fjords are composed, ground away? Really, it is unnecessary to give an answer. It answers itself. Some glaciers are rather broad, but still they have a limit, and so had these ancient glaciers, whose bed these fjords were ; and I suppose that though the "soft sand-stone, coal, black-lead," &c., which lay in the way, was worn away and floated seaward by the sub-glacial stream ; still, when the glacier reached its limits, what did not come within the area of the action of ice would remain. I believe this does not require a very great tension of the scientific imagination to conceive, and that even my opponent will acknowledge. Mr. Tayler in his, on the whole, short but admirably conscientious description of the Greenland fjords, mentions a fact in support of my theory, viz., that on the rocks on either side of these fjords are ice-markings. I would like him to explain these. It ought, however, to be mentioned that, except where the walls of the fjord are composed of trap, gneiss, or some other hard rock, we must not expect to see many marks of the grooving of the ice which formerly rubbed against them. For the action of the weather, disintegrating the surface of the rocks, or tumbling down huge masses into the sea, frost riving the rocks asunder, as well as the masses which in former times must have fallen on the side of the glacier in the form of lateral moraine, must have all helped greatly to efface any ice-markings which might have been formed. Both, however, Mr. Tayler here, and Dr. Rae in some remarks he made at the meeting in support of my views, mention seeing these markings, as I have seen them, both on the sides of these Greenland and Arctic fjords, and in other parts of the world. Mr. Tayler has presented a geological puzzle for my consideration in the form

of a Greenland fjord, and asked me to explain its formation on the
theory I have advocated. I daresay it would admit of a very simple
explanation, were we put in possession of all the facts in connection
with it. But, as Mr. Tayler's description is so meagre, until I
have seen it myself it would only be mere guess-work to attempt
showing its mode of formation. I do not advocate that everything
in the shape of an inlet of the sea was formed as I have mentioned.
On the contrary, doubtless, many inlets now classed under the name
of fjords were originally rifts and chasms in the country from
almost primeval times. It would be damaging to any theory to
claim for it the merit of explaining every fact of this nature ;
and it is scarcely fair to adduce some supposed exception to the
law enunciated, and thereby attempt to throw overboard all
the numerous facts adduced which prove that in the vast preponderance of typical cases it holds true. The glacier-bed theory
of fjords is a general theory applied to, and applicable to, all parts
of the world where fjords are found ; so that because seemingly
some glen in Greenland, or elsewhere, *looks like* an exception, it
must not be thrown aside. With, however, even less display of
ingenuity than has been exerted on throwing it in the way of my
theory, it could be accounted for, yet for the reason mentioned I
will not attempt this, but leave it to the opponents of the theory to
extract from it whatever comfort it is capable of affording them in
the way of argument.

5. *Volcanic Theory of the Formation of Fjords.*—What explanation the opponents of this glacier-bed theory would adduce is,
of course, not difficult to suppose. That fjords were formed by
the great volcanic agencies which in former times dislocated
the earth's crust is naturally their theory. Mr. Tayler has even
invented an hypothesis so ingeniously mechanical that I hope
he is not to be taken as a recognised exponent of the doctrines
of his school ! "It appears," he remarks,[1] "that at the time
of the elevation of the west coast of Greenland, a chain of
mountains about 50 miles in breadth, running nearly north and
south, was acted on in a wave-like manner, *i.e.* leaving depressions nearly equal to the elevations, and more or less at right
angles with the direction of the chain. These depressions, or
long valleys into which the sea runs, constitute the fjords,"
and so on. I am afraid this theory is *much* too ingenious to
be accepted by those who know anything of fjords or of igneous
action ; nor, I fear, is the general volcanic theory, though supported

---

[1] *Op. cit.*, vol. v. p. 90.

by illustrious names, so well founded as to be unassailable. It is,
to say the least of it, very remarkable, if fjords were owing to volcanic action, that they are not, as we might expect, found in countries
where there has been the most remarkable display of igneous
agency, or in countries where volcanic agency is equally well marked
with those countries in which fjords are found. On the contrary,
fjords are only found in northern and southern latitudes, where
glaciers either now form or could have formed, and nowhere
else; so that they must in some way be connected with climatal
agencies. Again, these fjords are only in the line towards the
sea, and always end at the shore, as if the agent which formed
them had been like a glacier making its way to the sea. A
volcanic rift is entirely different, and would never have shown
such a steady, uniform system of openings in the earth's surface.
If some great subterranean force had formed these openings in
the earth's crust we might have expected to find them on flats,
in mountains, in sandy tracts, in fact, anywhere—for a great
subterranean force would have risen the crust of the earth without
regard to locality. The fjords we *always* find surrounded by
mountains. Much more could be said, but it would be a mere waste
of time; for it must already be evident that, whatever agency has
formed these fjords, volcanic agency alone is not the one. I, with
all deference to my distinguished opponents, still think that the
glacier-bed theory is not untenable, in default of a better.

6. *Ramsay, Dana, Geikie, and Murphy, on Fjords.*—The views I have
enunciated, both here and elsewhere, regarding the action of Arctic
glaciers and glacier fjords, were first suggested to me when visiting
both sides of Baffin Bay and Davis Strait as early as 1861. I afterwards thought a great deal on the subject, during some years of a
lonely life, far away from scientific works or intercourse, while exploring the wild fjord-indented shores of British Columbia and Vancouver
Island.[2] Afterwards I saw enough in Greenland and Norway to
convince me that my early ideas had the germ of truth in them, and
that former writers, who attributed the formation of these to volcanic rifts alone, were not on the right track. I have accordingly,
in the course of the foregoing remarks, occasionally styled this
"my theory," for, until recently, I was unaware that the idea
had ever suggested itself to any one else. Though to me it is
a matter of perfect indifference who was the author of it, so
long as it is founded on truth, yet, in case it might be supposed

---

[2] See my "Das Innere der Vancouver Insel" (with map) in Petermann's
'Geographische Mittheilungen,' 1869, and 'Vancouver Island Explorations'
(V. I. Colonial Blue-book, 1865).

that I am adopting other men's ideas, I hasten to say I have recently learned that, without exactly explaining the formation of fjords as I have done, both Professors Dana and Ramsay had some years previously hinted at a similar explanation ; and more recently, Dr. Archibald Geikie, Murchison Professor of Geology in the University of Edinburgh, and Director of the Geological Survey of Scotland, has suggested that possibly the "lochs" on the west coast of Scotland might be so accounted for. Though none of these gentlemen took exactly the same view as I have done, or gave it such a general application, yet I am glad to have the support of men so able as they. I may be, therefore, excused if I add them as supporters of the glacier-bed theory of fjords.

Professor Ramsay [1] says : " Furthermore, as the glacialated sides and bottoms of the Norwegian fjords and of the salt-water lochs of Scotland seem to prove, each of these arms of the sea is only the prolongation of a valley down which a glacier flowed, and was itself filled with a glacier. . . . In parts of Scotland, some of these lochs being deeper in places than the neighbouring sea, I incline to attribute this depth to the grinding power of the ice that of old flowed down the valleys, when, possibly, the land may have been higher than now." Professor Geikie gives utterance to very similar views.[2] More recently still, Mr. J. Murphy, in a paper read to the Geological Society,[3] some months after mine was read to the Royal Geographical Society, apparently in entire ignorance of the writings of his predecessors, gives utterance to views even more decided regarding the part glaciers have played in the formation of fjords. His words are worth quoting : " Not many coasts in the world are cut up into fjords ; and nearly all that are so are western coasts in high latitudes. The fjord-formation is found in North-Western Europe, including Norway, the West of Scotland, and the West of Ireland ; in North America from Vancouver Island northward, and in South America from the Island of Chiloe southward. From Vancouver Island to Chiloe is an immense stretch of nearly straight coast-line ; but, at these limits, its character changes quite abruptly. The transition from straight to indented coast-lines coincides pretty equally with that from dry to moist climates ; and the change from the dry climate of Chili to the moist one of Western Patagonia is accompanied, as we might expect, by a depression of the snow-line on the Andes. It is now generally believed that the prevalence of lakes in high latitudes is,

[1] *Op. cit.*, vol. xviii. p. 203.
[2] ' Scenery of Scotland,' pp. 127, 183, &c.
[3] ' Quarterly Journal of the Geological Society,' vol. xxv. p. 354.

in some way, a result of glacial action : *it can scarcely be doubted that this is equally true of fjords,* and the coasts I have mentioned are those on which glacial action must necessarily be the most energetic : because west coasts in high latitudes are exposed to west winds (Maury's ' countertrades '), which deposit on the mountains in snow the moisture they have taken up from the sea."

7. *Nordenskjöld on Fjords.*—There is no scientific man living better acquainted with the varied phenomena of Arctic ice-action and Physical Geography than Professor Nordenskjöld, and these are his words, speaking of the Greenland shore :[1]— "The deep fjords evidently scooped out by glaciers."

I do not pit these authorities against the opponents of the glacier-bed theory of fjords; but only to show that, in supposing that glaciers and fjords have an intimate connection, I am not alone, as might be supposed from merely reading the arguments brought against my paper in this Society's ' Journal.'

9. The Northern Termination of Greenland.

What will be found to be the northern termination of Greenland is one of those geographical problems which, like the more trivial question of " What songs the Sirens sang," though a subject of legitimate speculation, is yet at the same time a matter which can only be settled by an Expedition like the one now preparing. Dr. Petermann has hazarded the opinion that Greenland stretches across the Pole and joins Wrangel Land north of Behring Strait. Without being able to express any decided opinion *pro* or *con.,* this hypothesis of the illustrious German geographer, except that it is just as reasonable as any other—but not more so—and as ingenious as is everything which emanates from the mind of my excellent friend, I think that recent discoveries point to the northern termination being somewhat different. Most likely it will be found that Greenland will end in a broken series of islands forming a Polar archipelago. That the continent (?) is itself a series of such islands and islets—consolidated by means of the inland ice—I have already shown to be highly probable, if not absolutely certain, as Giesecke and Scoresby affirmed (p. 25). It is not likely that the northern portion will be widely different.

The farthest view we have as yet had of it points to a group of broken islets. The open sea, or sea at least without any continuous

---

[1] ' Redogörelse för en Expedition till Grönland, År 1870 ' (Oversigt af K. Vet.-Akad. Förh. 1870, No. 10), and trans. ' Geol. Magazine, 1872,' p. 301.

or extensive floes, would seem to show that there is no narrow strait which would prevent the sea being cleared of ice in that direction. Farragut Point, and the other headlands which figure dimly on the map of the *Polaris* expedition, are probably capes of such islands. Nowhere in the Arctic Ocean have we found great unbroken stretches of land, and Greenland will most probably prove no exception to the rule. That huge glaciers, like the Humboldt glacier or those of Melville Bay, do not form the northern wall appear to me almost certain from the following facts: High up on the Greenland shore of Smith Sound we find the musk-ox (*Ovibos moschatus*); but this large and essential Arctic mammal is perfectly unknown south of Wolstenholme Sound. The glaciers south of that point seem to have formed an impassable barrier to its further progress, for the little difference in climate could have but a small effect on its range. In the winter season any portion of Greenland is sufficiently cold for it, and Smith's Sound in the summer is not much colder than most of the other parts of the continent. There must be, therefore, some physical cause for its being confined to that portion of the Greenland coast. Now comes in another most remarkable fact. On Shannon Island and the vicinity, in 74° N.L. (several degrees southward of where it roams on the opposite coast), the German Expedition to East Greenland found the musk-ox in great abundance. Again, so far as we know, it is as perfectly unknown on the south-eastern Greenland shores as it is on the south-western. How did it come across, for across Greenland it must have come? It is an American animal, and is nowhere found in Arctic Europe or Asia. It could not have travelled 700 or 800 miles across the inland ice, for such a large animal, independently of other considerations, requires a large quantity of food, which it could not have obtained on that icy waste. It must necessarily have passed over on dry land, where willows or other dwarf Arctic plants, on which it subsists, could be found. It might easily travel short distances on the frozen ice from island to island, and thus double the northern termination of Greenland, and stretch down the east coast for some distance, until again it met with an impassable barrier to its southern progress.

Take one further zoological illustration—and these illustrations, though seemingly trivial in themselves, are yet of extreme zoo-geographical interest—tending to show that the Greenland land must end not far north of latitude 82° or 83°. In 1822, Scoresby discovered a lemming near Scoresby's Sound on the east coast, which was named *Mus Greenlandicus*. It is now known to be a climatic variety of the European species, viz., *Myodes torquatus*.

Scoresby's specimen remained for long unique in the Edinburgh Museum, until in 1869 and 1870 the German Expedition found it in abundance on the same coast. This fact was interesting in itself, for it is unknown in the region, so far as has been explored further to the south, and in all parts of the west coast of Greenland explored up to the date of the *Polaris* Expedition. However, that Expedition found it not at all uncommon on the shores of the most northern reaches of Smith Sound (or the continuation of the gulf which goes under that name). The variety appears the same as on the east coast, but different from the lemming of the western shores of Davis Strait and Baffin Bay, which is *Myodes hudsonius*. Again the question suggests itself, how has this animal found its way across Greenland to the east coast, or *vice versâ* ? That its route has not been across the inland ice we may consider certain ; we may be sure it has been where food and footing could be found. In its migrations it will most likely be found to have been a companion of the musk-ox. The European ermine (*Mustela erminea*, L.) was also found by the German Expedition on the north-eastern coast, but is quite unknown on the west. If it should be found in Smith Sound also, the fact would form another remarkable zoogeographical problem for the English Polar Expedition to solve.

Lastly, it is, I venture to suggest, probable, or at least not improbable, that the aborigines, who to a small number now, but at one time in greater numbers, inhabited the east coast of Greenland did not stretch up from Cape Farewell as colonists from the west coast, but doubled the northern end of the country from the Smith Sound region. Like the Smith Sound people, the east coast Eskimo seem to want the kayak ; and it would be an interesting point to compare the implements, &c., of the remnant of " Arctic Highlanders " now living in Smith Sound, with the Eskimo of the south-eastern coast, and with the remains which the German Expedition discovered in the graves, which are now the only representatives of the fur-clad hunters and fishers who once inhabited that part of the coast explored by these intrepid voyagers.

If it should be found that the Greenland coast trends on the east towards the west and on the west towards the east, as there is some ground for believing, it is just possible that the English Arctic Expedition might be able to double the northern extremity of the continent, more especially if a sea comparatively free of fixed ice (I will not venture to say "an open Polar sea ") be found to lave its northern shores. Once on the eastern shores of Greenland, the observations of Captain David Gray, a Peterhead whaler, who last summer penetrated through the Spitzbergen ice-stream, and found open water to the north,[1] would seem to point out that the course of the expedition would then be clear. Such a feat in geographical importance and naval enterprise would be only second to the doubling of the northern termination of America—in other words, to the discovery of the north-west passage as achieved by M'Clure.

10. DEBATEABLE POINTS REGARDING THE PHYSICAL STRUCTURE OF GREENLAND.

Attention need scarcely be called to the fundamental point of all, viz., the improvement of our knowledge of the geography of the coast-line; to that, no doubt, the main efforts of the Expedition will be devoted. We know, as has been shown, comparatively little of the interior, and even the few expeditions which have attempted to penetrate eastward have only reached a few miles from the coast. Are there any mountains in the interior?—a question which I have ventured, reasoning from the facts before us, to answer in the negative : but Dr. Rink, incomparably the greatest of all authorities on Greenland, is (p. 58) by no means so positive on this question ; perhaps he is right. What is the nature of the soil under the ice ? Is it of the same character as the boulder-clay of Britain? The many points which ought to be investigated under these heads will appear in the geological instructions or will be evident to the reader after perusing the section on the " Greenland Glaciers and Ice."

Has the ice an abrading power? This is almost perfectly certain ; yet some observers—and still more some theorists—have attempted to deny this. Make every examination of the raised beaches on the shores of Smith Sound, and try, if possible, to test the question whether the shores of Smith Sound are actually rising, or are falling like the southern coast. On this point Mr. James Geikie, of the Geological Survey of Scotland—a most competent authority on all questions touching glacial deposits—suggests to me that " it would be very interesting to have determined whether the raised beaches of Greenland give any indication of changes of climate, such as having been observed in these deposits in Spitzbergen. Great banks of *Mytilus edulis*, *Cyprina islandica*, and *Littorina littorea*, occur in that island, and now are even found living in the Spitzbergen sea. It is true that *Mytilus* is occasionally seen attached to algæ in these regions, but such rare birds are but poor representa-

[1] Petermann's 'Geographische Mittheilungen,' March 1875.

tives of the banks of the same shell which are met with in the
same island. Mr. Nathorst, of the Swedish Geological Survey,
tells me that in 1870 he examined these shell-banks, and found
one made up of *Mytilus* resting upon a scratched rock surface (now
far removed from any glacier), and the scratches ran parallel with
the fjord. The *Mytilus* still lives in Greenland, as does also *Cyprina
islandica*, but *Littorina littorea* does not. Heer notices these circumstances in his paper ' Die Miocene Flora und Fauna Spitzbergens.' [1]
It would be worth while, I think, for the naturalists attached to the
Arctic Expedition to examine any raised beaches they may come
across, with a view to discover whether the facts bear on the conclusions drawn by Swedish geologists, for it is difficult to believe
that a considerable change of climate could take place in Spitzbergen without also leaving traces in North Greenland." All
these questions are of deep philosophical interest, and to their
solution the members of this Expedition are invited to apply themselves. We have shown, and the other portions of this manual only
confirm the remark, that in Greenland there is still much for the
geographer to do, and that when an ancient mariner wrote, 200
years ago, that " Greenland is a country very farre Northward,
. . . the land wonderfull mountainous, the mountaines all the year
long full of yce and snow, the plaines in part bare in summer-time
. . . where growes neither tree nor hearbe . . . except scurvy-grass
and sorrell . . . the sea . . . as barren as the land, affording no
fish but whales, sea-horses, seals, and another small fish . . . and
thither there is a yearely fleet of English sent," [2] he only wrote in
accordance with the knowledge of his time—and time has not
confirmed honest Edward Pellham's dictum.

[1] 'Öfversigt af Kongl. Svenska Vet. Akad. Förhand.' Band. 8, N°. 7, p. 23.
[2] In reality, though after the fashion of his time styling it " Greenland," the
devout old mariner—first of that long line of English seamen who have had the
courage to winter in Spitzbergen, and the good fortune to come back to tell
the tale—was describing Spitzbergen; but the quotation is sufficiently apropos to
remain without any very strict geographical criticism.

II.

ON THE BEST MEANS OF REACHING THE POLE.

By Admiral Baron von WRANGEL.[1]

THE vast accumulation of ice—which covers the northern seas in
immense fields, high hills, and small islands—subjects the navigator in these waters to incessant danger and anxiety: to struggle
with the elements, to overcome obstacles, to be familiarised with
dangers—all this is so habitual to the seaman, that he is sometimes even dull without it. The continual, uniform, and quiet
navigation in the regions of the trade-winds excites in the sailor a
desire for change : he encounters a squall with joy, welcomes even
a storm in the seas beyond the tropics not without a certain
pleasure ; and, confident in his skill, in the activity and indefatigable energy and experience of his crew, in the strength of his
vessel and soundness of all her parts, he does not fear the terrible
powers which so often put to the trial all his patience and all his
coolness. Such being the ordinary feeling of the seamen, it is not
astonishing that the Frozen Ocean has long attracted the navigators of all nations, but in particular those of England—that country
which has an indisputable right to be regarded as the first of all
maritime nations. Without taking into consideration the great
number of whalers, who have carried on their trade among the
mountains of ice in the most remote latitudes of the Atlantic,
England has sent out fifty-eight distinct expeditions to discover a
shorter passage to the Pacific, either by the north-west or north-
east channel, from the time of John Cabot (1497) to George Back
(1836): not one of these has been crowned with complete success.
In all these enterprises, however, one common aim, not specified
in the instructions, has ever been kept in view ; and this aim
has been more or less attained by every successive attempt—the
maintenance of the spirit of enterprise and the support of a land-

[1] From the 'Journal of the Royal Geographical Society,' vol. xviii.

able national pride, in the attainment of the laurels of disinterested
exploits, for the advantage of science, trade, and navigation—the
true sources of power and glory to every maritime people.

When, after nearly three centuries and a half, scientific men,
and even navigators, were persuaded of the improbability of the
existence of a north-west or north-east passage to the Pacific,
practicable for trade, the evident aim for new enterprises was
transferred to the invisible point of the earth—the North Pole.
The expedition of Captain Buchan, and the fourth voyage of the
indefatigable Parry, were undertaken expressly with that view.

This question, supported by the celebrated Barrow, has been
again moved in England, and has resulted in the exchange of
opinions on this subject between navigators and scientific men.

Captain Sir William Edward Parry, in a letter, dated the 25th
of November, 1845, to Sir John Barrow, proposes in a short outline a new plan for the expedition. Following the principles
there traced, a party would not, he thinks, meet with any of the
difficulties encountered by Parry himself in the latitude of
82° 45′ N., or about 2° to the N. of the extreme point of Spitzbergen, which was the starting-point of the Polar Expedition.
Having unequivocally assigned as the chief causes of failure in
those attempts—to which, however, no others can be compared
with respect to the difficulties overcome—1st., the broken, uneven,
and spongy state of the ice, covered with snow; and 2ndly., the
drift of the whole mass of ice in a southerly direction—Captain
Parry proposes, in order to avoid these unfavourable circumstances,
that the ship employed in the projected expedition should winter
at the northern point of Spitzbergen, and the party particularly
designed for the attainment of the Pole should leave the vessel in
April. About 100 miles north of this point there should be previously prepared a store of provisions, so that the party, at the
commencement of its journey, should not be too heavily laden;
and about the time of its return, according to the reckoning of
Parry, in the course of May, there should be sent out another
detachment with provisions to meet it about 100 miles further
from the place where the ship is wintered. Captain Parry founds
his hopes of success on the supposition that, in April and May,
the party would proceed about 30 miles a-day along the ice, which
would then offer an immovable, solid, and unbroken surface. He
also thinks it advisable to provide the expedition with reindeer.

Finding it difficult to make these ideas of Captain Parry accord
with those which I entertain respecting the state of the ice and
the circumstances indispensable to success in travelling along its

surface, I beg leave to express my doubts, and submit my ideas on this subject.

Expeditions were undertaken in the years 1821, 1822, and 1823, in the Siberian Frozen Sea, from two points of departure, distant one from the other, in the direction of the parallel, more than 1000 miles, viz., from the mouths of the rivers Lena and Kolyma. These expeditions occupied an interval from about the end of February to the beginning of May (O.S.), and the state of the ice does not at all seem to have been such as Captain Parry supposes it to be, to the north of Spitzbergen, in the course of April and May (N.S.).

Lieutenant (now Rear-Admiral) Anjou was stopped by thin and broken ice moving in different directions, in

1821. April 5 (O.S.)	at the distance of 20 Italian miles from the nearest shore	N. of the Island
1822. March 22.	22 Italian miles	Kotelnoy.
„ April 14.	60 „	E. of New Siberia.
1823. Feb. 28.	59 „	N. of the islands at the mouths of the Lena.

The expedition commanded by the author, which took its departure from the mouth of the Kolyma, encountered the same impediments :—

In 1821. April 3,	at 120 Italian miles..	N. of the nearest shore.
„ 1822. „ 12,	at 160 „	
„ 1823. March 23,	at 90 „	

But on the 27th of March the masses of ice, which were separated from each other by large channels of open water, were driven about by the wind and threatened the voyagers with destruction.

My hypothesis is founded on the above facts, collected during a three years' navigation in a sea whose depth is not more than 22 fathoms, and which is, so to say, landlocked to the south by the Siberian coast, and there defended from the winds and waves over a space of 180° of the compass; whereas the sea on the meridian of Spitzbergen has a considerable depth, and is exposed to the swell of the whole Atlantic. Therefore I cannot concur in Captain Parry's hopes that the ice can be in a state favourable to the execution of a journey towards the north in April and May.

Captain Parry's calculations as to the possibility of advancing 30 miles a-day seem to imply the employment of reindeer, and would render it necessary to provide the expedition with those animals: we must, therefore, conclude that that officer expects to obtain the necessary rapidity by the assistance of reindeer. If I am warranted in this supposition, I must remark that reindeer are far from being capable of advancing over the uneven surface of the ice, and are besides too weak to carry heavy burdens.

Sir John Barrow, in his work 'Voyages of Discovery and Research within the Arctic Regions,' &c., publishes the above-mentioned letter of Captain Parry, disapproving, however, his proposed plan, and anticipates greater success in the enterprise by accomplishing it in small sailing-vessels, fitted with the Archimedian screw (like the ships *Erebus* and *Terror*), and steering northward on the meridian of Spitzbergen : in other words—Barrow proposes the repetition of the former attempts, notwithstanding their failure, expecting success from more favourable circumstances. But here a question is naturally suggested—may there not exist means of reaching the Pole other than those which have been hitherto resorted to—means not liable to the various inconveniences already encountered during the several expeditions undertaken from the coasts of Siberia towards the north upon the surface of the ice, and which must be encountered in proceeding on foot, as Captain Parry proposes ?

The last Siberian expeditions were executed in a particular kind of sledges, called "Narty," drawn by dogs. The expedition, undertaken from the mouth of the Kolyma, travelled in this manner in 1823 (from the 26th February to the 10th May) 1533 miles, of which the greater part was along the shore towards the island of Koluchin, seen by Captain Cook during his navigation in a north-west direction from Behring's Straits. We proceeded upon the ice along the shore very successfully, but as soon as we left it the difficulties and impediments increased. If the coast of Siberia had a direction parallel to the meridian, the Kolyma expedition would have travelled 11° of latitude in one direction and the same in returning ; therefore, if the point of departure had been the 79° of north latitude, the expedition might have reached the Pole and returned to its starting-point.

The utmost limits of the coast of Greenland towards the north remain yet unknown ; but the meridian direction of its mountains and coasts allows us to suppose that, in proceeding along them, it is possible to approach the Pole nearer than from any other direction or even to reach that point.

The northernmost point of Greenland, Smith's Sound, seen by Captain Ross, is in latitude 77° 55' N.; and in latitude 76° 29', and on the island Wolstenholme, there is a village of Esquimaux. Taking all this into consideration, my opinion may be expressed in the following plan :—The ship of the expedition should winter near the Esquimaux village, under the 77th parallel, on the western coast of Greenland. There should be previously despatched to this point, in a separate party, at least ten narty, with dogs, and active and courageous drivers; the latter the same, if possible, as were employed in the Siberian expeditions,[1] likewise stores and provisions in sufficient quantity. In autumn, as soon as the water freezes, the expedition should go to Smith's Sound, and from thence further towards the north. On arriving at 79°, it should seek on the coasts of Greenland, or in the valleys between the mountains, for a convenient place to deposit a part of the provisions.

In February the expedition might advance towards that place; and in the beginning of March another station, two degrees further north, might be established. From this last point the Polar detachment of the expedition would proceed during March over the ice, without leaving the coasts, keeping along the valleys, or on the ridge of mountains, as may be found most expedient, but deviating as little as possible from the line of the meridian, and shortening the distance by crossing the straits and bays. A part of the men, dogs, and provisions, should await their return at the last station.

The expedition, to reach the Pole and to return, must traverse in a direct line nearly 1200 miles, or, including all deviations, perhaps not above 1530 miles, which is very practicable, with well-constructed sledges, good dogs, and proper conductors.

If the most northern limits of Greenland, or the Archipelago of Greenland Islands, should be found at too great a distance from the Pole, and the attainment of that point seem impossible, the expedition might at any rate draw up the description of a country hitherto absolutely unexplored, and would, even by so being, render an important service to geography in general.

---

[1] The success of such an enterprise would chiefly depend on the kind of dogs, the experience and courage of the conductors, and the form of the sledges. It certainly will not advance rapidly if Esquimaux or Tchouktschi dogs are employed, because these are entirely unaccustomed to such long journeys; nor with Esquimaux or Tchouktschi drivers,—men without courage or activity.

III.

ON THE DISCOVERIES OF DR. E. K. KANE, U.S.A. (1853–55).

By Dr. Rink, Director of The Royal Greenland Board of Trade, and formerly Inspector in Greenland for the Danish Government.[1]

The author of the work above quoted makes the following remark in the Introduction : " This book is not a record of scientific investigations ;" and adds, that his aim has been to publish a narrative of the adventures of his fellow travellers, and that he has attempted very little else. Nevertheless, on perusing this promised " simple story " of a voyage, we find it embellished with scientific theories extending far beyond the bounds of such a narrative. As these speculations relate to a subject, the examination of which has occupied me during nine years, namely, the Physical Geography of Greenland, I feel called on to subject them to a somewhat closer inquiry. As his richly and elegantly illustrated work has awakened great sensation, nay even partly placed the other Polar expeditions in the shade, I am led to think that a communication of my views respecting this matter will not be entirely without interest to the Society.

It is well known that the active and undaunted American traveller, Dr. Kane, unfortunately so early carried off, attempted, in the year 1853, to go farther north up Smith Sound than Captain Inglefield, the year previously, had done ; but that he only succeeded in taking his ship a trifling distance farther than Inglefield ; that he was then frozen in, lost his ship, and in the year 1855 saved himself and party by returning, in boats, to the Danish colony of Upernivik. From his two years' winter quarters in Van Rensselaer Bay, on the east side of the Sound, he, by the help of dog and drag sledges, undertook expeditions in different directions, partly across to the American side, but mainly along the coast, pursuing it northward, to find, if possible, the northern end of Greenland.

---

[1] From the 'Journal of the Royal Geographical Society,' Vol. xxviii.

What was discovered on these tours must be regarded as the real profit of the expedition, and I will here confine myself to the two points which have cast the chief lustre over it. *First,* that which concerns the unknown interior of Greenland, the glaciers and floating icebergs that issue thence, about which the author expresses himself on occasion of having discovered a glacier on the coast of Greenland, between 79° and 80° N. lat., to which he has given the name of Humboldt. *Secondly,* a sledge expedition undertaken by Morton (one of the ship's crew who it seems was steward), in conjunction with Hans, a Greenlander from Fiskernaesset; whereby they are said to have come to the margin of an *open* sea, which is presumed to occupy the whole region around the North Pole, and to be kept open by a branch of the Gulf-stream; and, besides this, to have discovered the most northern lands on our globe, which, according to their description, are likewise laid down on the chart and called "Victoria and Albert," "Washington," &c., Lands.

As regards the first of these points, I must repeat what I have explained in my work on North Greenland,[1] namely, how the whole of the inner mainland, regarded from the outer land, appears buried under one uniform covering of ice, which sends its branches down into all deep fjords; how these branches are pushed down into the sea, and yield annually large masses of ice in the form of floating icebergs or calves. The glacier discovered by Kane, which he has named "the great glacier of Humboldt," and which has called forth much admiration, even in well-known geographical journals, has been represented as the crowning point of the discoveries made by the expedition, but which is really nothing more than what can be observed in the interior of most of the Greenland fjords, from the southernmost to the most northern reached point.

The reason why Kane has not had an opportunity to observe these, and that the one discovered by his expedition has therefore appeared to him so remarkable, lies in the simple fact, that such ice formations in general lie hid behind the numerous high islands and peninsulas, which almost form the outer coast of Greenland towards Davis Strait, and which, with regard to snow and ice, do not show any other phenomena than the higher parts of the mountain chains of Europe.

Now as the different discovery-ships, that have sailed in search of the North-West Passage and of Franklin, have always rapidly hurried through Davis Strait, and have only touched at one or other of the Danish colonies, it is no wonder that the numerous

---

[1] 'De Danske Handelsdistrikter i Nordgronland.'

remarkable ice-fjords, which require a longer time to travel through
and examine, have more or less escaped attention. Kane had thus
either not seen these ice formations, or only had an occasion of
seeing them from a great distance, before he came to the place
where he was frozen in and had to pass two winters. Humboldt
glacier does not even seem to belong to the most remarkable among
them, as even in the very southernmost of our Greenland districts,
at Julianehaab, we have opportunities of observing just as remarkable phenomena of this kind.

With respect to the *second* point—namely, the *Open Polar Sea*,
discovered by Morton the steward and the Greenlander Hans—
the manner in which Morton's journey is described by Petersen,[1]
the Dane, who accompanied the expedition as interpreter, seems
to give a clearer picture of its result than that which Kane has
sketched.

This discovery of an Open Sea gives Kane occasion to make a
comparison with other Polar expeditions, and he goes as far back
as the days of Barentz in 1596, and "without referring to the
earlier and more uncertain chronicles," he mentions the Dutch
whale fishers, Dr. Scoresby, Baron Wrangel, Captains Penny and
Inglefield, and shows how they have all spoken about large openings in the ice around the North Pole. He shows likewise how
these have all been found to be "illusory discoveries," and anticipates the objection that "his own may one day pass within the
same category" by extolling the far larger scale on which *his Open
Sea* has been observed. Petersen confines his remarks on this
subject to the following:—

"The Greenlander, Hans, was sent after them with the dog-sledge in order to
continue the journey still farther towards the N., and when he reached their
sledge (i. e. a drag-sledge that had been sent out earlier), he and the steward
Morton proceeded onwards. They reached the Sound of which the Esquimaux
had spoken. This Sound was open; probably cut up by the strong current
they had observed there. It was, however, Midsummer, so that the sun had
perhaps aided the current in getting away the ice. After this expedition no
other such was attempted." [2]

It is a known fact that, here and there under the coast of North
Greenland, places are found which, on account of the strong

[1] *See* vol. i. pp. 280-310. Petersen is a man well known to me. He was appointed foreman in the trading service at Upernivik. His communications bear
the full impression of truth, and are written in a clear and simple style, without
boasting and self-praise, although he has been of great service to the expeditions
that he accompanied as interpreter—*viz.* Penny's and Kane's. He is now serving
with Capt. M'Clintock.
[2] *See* 'Erindringer fra Polarlanderne,' p. 12.

current, do not freeze, even in the severest winter, although the
whole waters round them are covered with ice of two to four feet
thick, and Kane himself remarks, that in the most rigorous cold
he has found such stream-holes. As soon as the Spring commences
these stream-holes expand themselves, as the ice in their neighbourhood is always thinner and sooner thawed, either above, by the
sun, or below, by the under-current.

Now, as Morton's expedition was undertaken at Midsummer, and
as he found such an opening in the ice, not more than 90 miles from
the place where they, the year before, had been able to navigate
the vessel, and as there was an unusually strong current running
in this opening, which just appeared where the Strait became
smaller, nothing is more probable than that this opening was just
such a stream-hole, in which opinion I must concur with Petersen,
until stronger proofs be adduced in favour of the hypothesis of
an Open Polar Sea kept open by a branch of the Gulf-stream
deflected from Nova Zembla to the Pole: a solution of a problem
which has occupied Geographers since 1596, if not farther
back, &c., &c.[1]

Next, as to what concerns the lands that are said to surround
this enigmatical Sea with a coast of 90 to 130 miles in extent,
which Morton measured almost at a single glance, and which Kane
has been able to lay down on his chart, even with an exact coast
margin, adorned with celebrated names, and accompanied in the
text with correct statements of the heights of mountains (Mount
Parry, &c., &c.), I must express a well-founded doubt of the correctness of all this.

The ship, as stated, was frozen in on the coast of Greenland, in
78° 37′ N. lat., in the beginning of September, 1853. Of the expeditions that were sent out the same Autumn with boats or sledges,
one reached, as presumed, 79° 50′ N. lat. along the same coast. In
March, 1854, Dr. Kane sent out a sledge expedition, which was
obliged to return without result; the eight travellers who took
part in it were in the greatest danger of being frozen to death;
three of them had a foot or toes amputated, and one died a few
days after his return. Of the later expeditions, the one under Dr.
Hayes was directed towards the opposite, or American coast, which
he traversed to 79° 45′ N. lat. under great sufferings from snow-
blindness. The others kept under the coast of Greenland, and did
not get farther than Humboldt glacier, or about 79½° N. lat.; with
the exception only of the one undertaken by Morton and Hans,

[1] See Kane's 'Considerations,' vol. i. pp. 301-309.

who, according to their own statement, reached 81° 20′ N. lat., from
which point they supposed they had seen land as far as 82° 30′ N.
lat. ; these two members of the expedition alone came to the *Open
water.* The breadth of the whole of the northernmost part of
Baffin Bay, thus explored, was from 8 to 16 geographical miles
between the coasts of Greenland and America.[1]

After the first excursions in the vicinity of their winter quarters
attention was directly drawn to the great Humboldt glacier, and
Kane had an occasion, one clear day in April, to survey it closely ;
and then remarks :—

"My notes speak simply of the 'long ever-shining line of cliff, diminished
to a well-pointed wedge in the perspective ;' and again, of 'the face of glistening ice, sweeping in a long curve from the low interior, the facets in front
intensely illuminated by the sun.' But this line of cliff rose in a solid
glassy wall, 300 feet above the water-level, with an unknown, unfathomable
depth below it ; and its curved face, 60 miles in length from Cape Agassiz to
Cape Forbes, vanished into unknown space at not more than a single day's
railroad travel from the Pole. The interior with which it communicated, and
from which it issued, was an unsurveyed *mer de glace*, an ice-ocean, to the eye,
of boundless dimensions.

"It was in full sight—the mighty crystal bridge which connects the two
continents of America and Greenland. I say continents ; for Greenland,
however insulated it may ultimately prove to be, is in mass strictly continental.
The least possible axis, measured from Cape Farewell to the line of this glacier,
in the neighbourhood of the 80th parallel, gives a length of more than 1200
miles, not materially less than that of Australia from its northern to its
southern Cape.

"Imagine, now, the centre of such a continent, occupied through nearly its
whole extent by a deep unbroken sea of ice, that gathers perennial increase
from the waterparting of vast snow-covered mountains, and all the precipitations
of the atmosphere upon its own surface. Imagine this, moving onward like a
great glacial river, seeking outlets at every fjord and valley, rolling icy cataracts into the Atlantic and Greenland seas ; and, having at last reached the
northern limit of the land that has borne it up, pouring out a mighty frozen
torrent into unknown Arctic space.

"It is thus, and only thus, that we must form a just conception of a
phenomenon like this great glacier. I had looked in my own mind for such
an appearance, should I ever be fortunate enough to reach the northern coast
of Greenland. But now that it was before me, I could hardly realize it.

"I had recognized, in my quiet library at home, the beautiful analogies
which Forbes and Studer have developed between the glacier and the river.
But I could not comprehend at first this complete substitution of ice for
water. It was slowly that the conviction dawned on me, that I was looking
on the counterpart of the great river-system of Arctic Asia and America.
Yet, here were no water-feeders from the south. Every particle of moisture

[1] *See* vol. i. pp. 225-228.

had its origin within the Polar circle, and had been converted into ice. There were no vast alluvions, no forest or animal traces borne down by liquid torrents. Here was a plastic, moving, semi-solid mass, obliterating life, swallowing rocks and islands, and ploughing its way with irresistible march through the crust of an investing sea."

As Kane, in this section of his work, just expatiates upon the nature and quality of the whole of Greenland and its unknown interior, it is chiefly at this place that I must refer to my previously cited work ; in the first section of which, at page 10, I have treated on the extension of the land-ice, and the origin of the floating icebergs. But as the subject is rather comprehensive, I will here confine myself to the following remarks :—

The interior, with which the glacier stood in connection was : "an ice-ocean, to the eye, of boundless dimensions." That this ice-ocean could not be overlooked at that place certainly does not signify much with regard to its extent ; but farther on, he remarked that it occupies the whole centre of Greenland, right down to Cape Farewell. Now, from what source does the author know this, as he only cites a few places, quite in the neighbourhood of his winter harbour, where he has followed the margin of the inland ice, and had never been in the fjords of Greenland, between Upernivik and Cape Farewell ? I for my part have employed eight years in examining to what degree the interior was covered with ice, by pursuing it from fjord to fjord ; and nevertheless I have been obliged to confine myself to conjecture with regard to many extensive tracts that lie between these fjords; and my own explorations in this direction, must, as we shall see, be supposed to have been unknown to him. In the account of his first voyage,[1] he says of the Omenak fjord, that he could see into its mouth whilst sailing up the Strait ; that its interior had never yet been explored, and that there was great probability that it passed right through the country to the Atlantic Ocean. But if we admit this central ice-ocean as existing, what does it then signify ? that this ice-ocean moves like a great ice-river (from south to north ?), rolling cataracts of ice out to both sides in the Atlantic and Greenland seas, until it reaches the northern boundary of the country, and there pours forth a mighty frozen stream, Humboldt glacier, in that unknown Arctic space ? I cannot follow the author in his bold flight over the icy desert of Greenland, and still less can I conceive that he, in all this, only sees a confirmation of what he had already earlier foreseen in his own mind, if he " *should ever be fortunate enough to reach the northern coast of Green-*

---

[1] 'Grinnell Expedition,' 1851, p. 53.

land,"—that which he presumes to have discovered on this expedition. The reality is, that wherever one attempts to proceed up
the fjords of Greenland, the interior appears covered with ice ; but
there is no reason whatever to assume that this applies to the
central part of the country, in which one, on the contrary, just as
well may assume that there are high mountain-chains, which protrude partly from the ice. A remarkable movement is found in this
ice-mass ; but this is so far from having a kind of main direction
after the central axis of the land towards the Humboldt glacier,
that this arm of the ice, on the contrary, seems to belong to those
that are in a less degree of motion, whereas the greatest agency
takes place around Jakobs-havn ice fjord, Omenak fjord, and others.
Farther, this movement can only be measured by the masses of ice
that pass annually out of these fjords, and of which one can only
obtain a tolerable conception by remaining for a long time at the
mouths of the fjords. These ice-fjords point out probably the rivers
of the original land, now buried under ice. Whereas no conclusion
can be drawn from the ice itself and the appearance of its branches
that go down to the sea, for it is almost quite uniform everywhere
from Julianehaab to Upernivik.

The author, in concluding his remarks, says it was first when he
saw Humboldt glacier that Forbes's and Studer's idea of the like-
ness between the glacier and the river began slowly to dawn on
him ; but the same species of glacier, which these celebrated naturalists have examined on the Alps and in Norway, is found in many
places on outer-Greenland, or what I would call ice-free Greenland.
These Kane had seen at Disko, near Upernivik, and other places,
before he reached " Humboldt glacier." In order to examine its
significance in comparison with the rest of the branches of inland
ice, he must have made observations and calculations of how many
icebergs it annually yielded to the sea, as from its appearance he
could scarcely form any opinion. By seeing such a branch of inland ice, on account of the uniform ice-plateau whence it issues, one
gets a smaller impression of its similarity with a river than by
seeing the Alpine glaciers and the glaciers on the outer coast of
Greenland, as these just fill up clifts which—to judge from their
form—must be beds of watercourses. Those arms of inland ice,
which send scarcely any ice into the sea, show, on the contrary,
about the same appearance as those that send out annually thousands
of millions of cubic feet of ice into the sea, and therefore must be
supposed to be maintained by river territories of many hundred
geographical square miles.

I now proceed to examine its signification as a sort of connecting

link between Greenland and the American continent. Dr. Kane
says "it was in full sight—the mighty crystal bridge which connects the two continents of America and Greenland;" and afterwards, in a note, "I have spoken of Humboldt glacier as connecting
the two continents of America and Greenland. The expression
requires explanation," &c. Difficult as it is to understand, Dr.
Kane seems to mean that Greenland is separated from, and therefore
half connected with, the Arctic-American Archipelago by a less broad
Sound, beyond Humboldt glacier.

Petersen says, that Kane himself would have undertaken an
excursion to the north in the middle of April 1855, but that he
could not get the Esquimaux to accompany him, as they would
only go bear-hunting around the ice cliffs near Humboldt glacier,
and thus Kane was only absent 24 hours on this tour. Kane
says that as he could not reach the Open Water, he sought compensation in a closer examination of the great glacier, of which he now
again takes occasion to give a lively description, concluding with
the following allusion to the previously-mentioned idea of the connection between Greenland and America:—

"Thus diversified in its aspect, it stretches to the north till it bounds upon
the new land of Washington, cementing into one the Greenland of the Scandinavian Vikings and the America of Columbus."

In the earlier sections there is spoken of the extension and movement of the inland ice; here is specially mentioned the manner in
which the floating icebergs tear themselves loose from that side
which goes out to the sea—the *calvings* as they are called in the
ice-fjords. None of those engaged in the expedition had had an
opportunity to make direct observations in these respects. In order
to obtain the necessary prospect, Kane climbed up "one of the
highest icebergs," whilst his fellow-travellers rested themselves.
From here he meant he could see that

"The indication of a great propelling agency seemed to be just commencing
at the time I was observing it."

It appeared to him as if the split-off lines of the fast land ice, which
signify the beginning of the loosening, were evidently about to
extend themselves. As the *calving*, however, did not follow, Kane
confines himself to remark respecting it—

"Regarded upon a large scale, I am satisfied that the iceberg is not disen-
gaged by *debâcle*, as I once supposed. So far from falling into the sea, broken
by its weight from the parent glacier, it rises from the sea."

He next adds that

"The idea of icebergs being discharged, so universal among systematic

writers and so recently admitted by himself, seems now to him at variance with
the regulated and progressive actions of nature."

By this I conclude that Dr. Kane had not seen my work on North
Greenland, or, at all events, that part of it which treats of the extension of the land-ice and the origin of the floating icebergs, and
wherein it states—

"But from what has been already mentioned, it must be evident that the
icebergs must not be considered as breaking loose and falling down from
precipices ; one might rather say that they lift themselves," &c. &c.

That Kane did not know this is certainly very striking to me, as
the literature which treats of the glaciers of the Polar lands, and
especially those of Greenland and the origin of the icebergs, is not
great.   Dr. Kane had sought information respecting the nature of
the country in our Danish colonies, and as my above-mentioned
work is cited in his own, if not by himself, still by his assistant,
Charles Schott, in the Appendix XIII., p. 426.[1]   He says also, at
page 150, that the height of the ice-wall at the nearest point was
about 300 feet, measured from the water's edge.   As a consequence
thereof the floating icebergs, which lay before it and were detached
from this ice-wall, must have been, on the average, above 300 feet,
if they should be imagined as formed by an elevation during the
time of being detached.

I have accurately measured many frozen icebergs, particularly in
the winter, on Omenak fjord, and I have thereby come to the result
that the common height of the larger ones, and especially of those
that may be supposed to lie, in some measure, in the original position
which they had had after their breaking loose, was somewhat more
than 100 feet.   I have also measured them as high as 150 feet, and
I have seen some that I should estimate at 200 feet high ; but this
was when there were points or edges that had come to jut upwards
by the mighty ice-block having turned and changed its position in
the water.   That the whole of the collected mass of icebergs before
the Humboldt glacier should have been considerably more than
300 feet in height generally—the highest, consequently, even 500
feet—I can certainly not disprove ; but I must strongly doubt.

We now come[2] to the remarkable sledge expedition of Morton
and Hans, on which they first passed the whole exterior margin of
the great glacier, with the icebergs lying before. and those torn
from it and floating about ; they then drove farther towards the
north, found the ice more and more unsafe, and were at last inter-

---

[1] See also 'Journal of Royal Geographical Society,' vol. xxiii. p. 145.—ED.
[2] See vol. i. pp. 280-310 of Kane's Work.

rupted by the Open Sea, when they drove some distance along the
shore, and lastly Morton went alone on foot as far as he could to
obtain a survey of the navigable water farther towards the north.
The whole journey, from the moment they saw the Open Sea until
they were compelled to return, after a very difficult passage, during
which they were also bear-hunting, lasted only three days, or from
the 21st to the 24th of June.

What Morton saw in these three days is the foundation for the
whole theory of Kane's Open Polar Sea, and whatever stands in
connection therewith. Kane gives us this account with his own
explanations, and in a separate Appendix he has communicated
Morton's own journal. It is stated that this man had instruments
with him to determine the geographical positions. As far as I can
judge from the chart, as laid down in the first volume, and from the
Appendix, No. VI.,[1] more than 20 points of longitude and latitude
are determined by him on that toilsome journey beyond the Humboldt glacier, besides the numerous points on the opposite coast, to
which they did not come, and which, therefore, appear to be laid
down only after bearings.

When I consider the great haste required to reach the farthermost point towards the north, and to return before the ice broke up,
the very difficult and toilsome passage through deep snow, over
openings, the most trackless ice-walls, &c. &c., I cannot sufficiently
admire Morton's dexterity in attending at the same time to these
observations which require so much repose and accuracy.

The travellers drove past the floating icebergs that were torn
loose from the glacier and lay piled up before it. Several reasons
are adduced to show that it could be ascertained that they were
formed or torn loose very recently, as they had a fresh shining
surface and no projecting foot under the water. It is, however,
especially from the accounts given of this place that I conclude
that the Humboldt glacier does not belong to the most active of
the inland ice-streams of North Greenland. The icebergs lay only
a few Danish miles out from the fast land-ice, and one must consider that they have perhaps taken several years to be filled up,
as all the navigable waters thereabout were frozen ; they could
scarcely come out any other way than towards the south, and this
passage perhaps opens only now and then in different years. The
great ice-fjords that are known in North Greenland are annually
cleared of great masses of ice, that are driven to sea. If this were
not the case, the inner navigable waters would soon be stopped up,

---

[1] The *astronomical* observations obtained by Morton are three meridional
altitudes of the sun.—ED.

and the incessantly-propelled land-ice extend itself over the surrounding land.

After having passed the icebergs, they came to the place where
the sea-ice on which they drove became thinner and thinner, so
that the dogs trembled, and at last they durst not drive farther on
it, but sought the land, or rather the firmer ice-edge that lay immediately along the shore.   At last the ice gave place for quite open
water, and here it is stated, at page 288, that—

"The tide was running very fast; the ice-pieces of heaviest draught floated
by nearly as fast as the ordinary walk of a man, and the surface pieces passed
them much faster, at least four knots."

Kane has already given an excellent description of a stream-hole ;
but had it been the margin of the Open Sea moved by the swell,
the ice would have kept its thickness, at least to some extent, just
as one approached it, but it would have been broken, screwed up,
and thus more or less in drift.   In short, such a margin of ice is
cut off sharper, with respect to thickness, whereas a successive
transition from ice to water is found around a stream-hole, *for which
reason it is so dangerous to approach such places.*   The above-mentioned
tide-stream of four knots is even so strong, that one (particularly as
it was in a pretty large sound, and not in a narrow pass of some few
yards in breadth) can already conclude that in such a place no ice
would be able to hold in the month of June, even to a considerable
circumference.   Even farther up Morton observed that the ice-pieces
drifted at the rate of four miles an hour, and that the stream varied
first from north to south and then from south to north, just as is
the case everywhere in the inner navigable waters along the coast
of Greenland, originating from the ebb and flood. (*See* vol. ii., p.
376.)

The last-mentioned observation was made by Morton on the 22nd
of June, consequently there was not until that moment the most
remote reason to suppose an Open Polar Sea.   The Sound had likewise a direction north, and there was thus no sign whatever that
the coast under which they found themselves turned towards the
east, or that they found themselves *at the end of Greenland.*   We
will now consider the adventures of the two following days, after
Morton's own description (vol. ii., pp. 377, 378).   These adventures
form the main foundation for the ideas about the end of Greenland
—the Open Polar Sea—the Gulf-stream, which warms up the Pole
—the solution of that problem which has occupied the geographical
world since 1596, &c. &c. ; and with these must stand or fall the
whole of that splendid building, of which Kane has sketched a
drawing in vol. i., pp. 301-309.

On the 23rd of June Morton and Hans started, but not before noon, in consequence of a continued gale from the north, but after driving about 6 English miles they found the ice along the coast quite broken up and impassable. They therefore made a halt with the sledge, and undertook a journey on foot, but returned and encamped by the sledge.

The following day, the 24th of June, they started on foot very early in the morning; their intention was to come past a high cape, behind which there was still hope that they could get a free prospect towards the east, and thus see the end of Greenland. After a very toilsome wandering, as they were sometimes obliged to crawl over cliffs and sometimes to spring over loose floating pieces of ice, they fell in with a she-bear and her cub, which they killed, and then boiled a strengthening dish of the flesh on the spot, as they found some plants and a piece of a sledge, whereof they made a fire. As yet nothing was discovered that could lay the foundation to the above-named theories, and nevertheless all was to be attained before the following day. On account of the importance of the events that occurred between, I will give Morton's statement, as it will be found in the place cited :—

"After this delay (the bear-hunting) we started in the hope of being able to reach the cape to the north of us. At the very lower end of the bay there was still a little old fast ice over which we went without following the curve of the bay up the fjord, which shortened our distance considerably. Hans became tired, and I sent him more inland where the travelling was less laborious. As I proceeded towards the cape ahead of me the water came again close in-shore. I endeavoured to reach it, but found this extremely difficult, as there were piles of broken rocks rising on the cliffs in many places to the height of 100 feet. The cliffs above these were perpendicular, and nearly 2000 feet high. I climbed over the rubbish, but beyond it the sea was washing the foot of the cliffs, and, as there were no ledges, it was impossible for me to advance another foot. I was much disappointed, because one hour's travel would have brought me round the cape. The knob to which I climbed was over 500 feet in height, and from it there was not a speck of ice to be seen. As far as I could discern the sea was open, a swell coming in from the north-ward and running crosswise, as if with a small eastern set. The wind was due north—enough of it to make white caps—and the surf broke in on the rocks below in regular breakers. The sky to the north-west was of dark rain-cloud, the first that I had seen since the brig was frozen up. Ivory gulls were nesting in the rocks above me, and out to sea were mollemoke and silver-backed gulls. The ducks had not been seen north of the first island of the channel, but petrel and gulls hung about the waves near the coast."

"June 25.—As it was impossible to get round the cape I retraced my steps," &c. &c.

With this, the exploration of the open Polar Sea,[1] and the farthest
lands on our globe, was ended. Morton felt himself disappointed
in not being able to come past that terrible cape, which hid his
prospect towards the east. I, for my part, was not disappointed on
reading that such a hindrance arose before him. I know it from
sad experience, as I, during three consecutive winters, have followed
the winding coasts of North Greenland in dog sledges, in order to
lay them down on my chart. I know these bewitched points which
continue to shoot forth when one thinks one is at the end of an
island, these endless promontories which one must get past before
one can reach the right promontory, and can turn round ; these hills
—these eternal tops—that shoot up when one ascends the cliffs,
before one reaches the right top, whence one can have the wished-
for prospect. I have passed half a day thus only to get the wished-for
general view over one single fjord-arm, and that even sometimes in
vain. What must it then not be, when one on an afternoon, and on
foot, seeks to reach the unknown end, to use Kane's own words, of
a " whole little Continent ? "

We will now return to Kane's representation, and, on account of
its considerable extent, confine ourselves to inquire into the most
important conclusions, through which he comes to such great results
from the facts communicated above.

Dr. Kane remarks in several places, that although it blew a strong
and almost stormy north wind during those days when Morton travelled along the open water, there came only some few half-dissolved
pieces of ice drifting from the north, and at last none at all. This
shows, if one will draw any conclusion whatever from it, that the
navigable water, a good way from the mouth of the narrow pass, in
which the stream was so extremely rapid, had been covered with
still good winter ice. For if it were really on the border of the open
sea one might expect to find much loose drift-ice between the
margin of the fast ice over which they had driven, and the quite
open sea ; and there was a great probability that such drift-ice
must appear and press on during a continued north wind. A
sudden beginning of a perfectly ice-free sea is scarcely to be
imagined.

[1] With reference to the latitude of the northernmost point reached by Morton,
he states in his Journal, p. 378, vol. ii., " We arrived at our camp where we had
left the sledge at 5 P.M., having been absent 36 hours, during which time we had
travelled twenty miles due north of it. June 26th.—Before starting I took a
meridian altitude of the sun." This observation is worked at page 388 in the
same volume, where the result appears as .. .. .. 80° 20' 2"
Add 20 miles according to the above remark 20 0

Latitude of the farthest point reached by Morton .. 80 40 2

An important criterion, whereby to judge if one has open water, is the *ground swell* of the sea. This is seen at Julianehaab, when the ice from the east coast is expected in the spring. To look after the ice itself from hills of some hundred feet in height is not of much use, for if it be first in sight it is also very near, and in a short time is on land. But in general one can know its proximity by the cessation of the ground-swell several days beforehand. To observe this with certainty the weather must be quite still, for the swell which even a common wind produces makes the observation uncertain. Kane adduces the swell and surge as proofs of the Open Polar Sea; but as it is expressly stated that *it blew almost a storm the whole time*, the effects of such a storm on an open surface of the sea, of possibly 20 or 30 miles in extent, are sufficient to make the presumed observation perfectly invalid. Still more uncertain does the observation of Morton appear to me, that the swell caused by the wind from the north, which he pretends to have remarked from the farthermost point of land, was acted on by another swell from the east, behind that Cape which concealed the end of Greenland and the beginning of the great Polar Sea from his view.

A third fact which Kane adduces in favour of his theory of the Polar Sea, is the increasing abundance of animals and plants in the district to the north of the glacier. It is mentioned in particular that seals and sea-fowl were seen in great numbers in, as well as around the neighbourhood of, the open water. Passing over the more cursorily touched observation, that the birds flew in an eastern direction behind the oft-mentioned cape which Morton could not come past, I shall only remark that I, on the contrary, regard that flocking together of sea animals and birds *as a sign of one single opening in the sea*, the rest of which was covered with ice. Such openings are just characteristic gathering-places for seals and sea-fowl. Nor do the plants which the Greenlander Hans is said to have seen, but no specimens of which were collected, and which from his bare description, are determined and inserted with Latin names of their genera and species at page 462, appear to afford any weighty proof of the Open Sea and an increasing mildness of climate towards the North Pole.

I now come to the real question, the *knob* to which Morton climbed when he could not come farther, and from which he, " as far as he could discern," found the sea *Open*. He says that it was over 500 feet in height, though he likewise remarks that the cliffs around, to a height of 100 feet, which were difficult to reach, were quite perpendicular. As far as I can make out, this is the same point to which Kane, at page 299, gives a height of 300 feet; at page 305, of 480 feet; and lastly, at page 307, where he compares it with the points from which former expeditions are supposed to have seen the open sea, of 580 feet. How this very doubtful height was measured, is not mentioned, and yet it is from this position that the size of the surveyed *open* space is to be given. Nor have I been able to find due information of how clear the air was, nor where the sun was at that time. Morton speaks of a dark rain-cloud in the N.W.; and a delineation of the open sea, with Morton in the foreground, "*from description*," as it is called, is also given at page 307. But with the exception of a mysterious round body bathing one half in the sea, but which cannot be the sun at this season of the year, a long way above the horizon, even at midnight, one sees nothing but the sea bounds bordering the horizon. Neither is it quite clear in what direction the oft-mentioned Cape concealed the prospect towards the east. We see the coast-line on the chart broken abruptly off by the farthest point that Morton saw. We ought to have the necessary information about all these questions in order to judge of the correctness of the calculations by which Kane, at page 302, came to the result, that Morton could see from his "look-out" to a distance of 36 miles, and that he had consequently sur-veyed an Open Sea of more than 4000 square miles. Every one acquainted with the nature of "*looking out*" after ice will admit the folly of determining with certainty, by sight alone, from a height of some few hundred feet, that flat ice is not to be found on the sea in the farthest margin of the horizon, or at a distance of 36 miles. If even, as I much doubt, it could be possible, under very favourable circumstances, to discover it at such a distance if it were there, it however becomes an impossibility to determine its absence with certainty. If we now remember that the part of the sea which Morton had already passed, after he left the Humboldt glacier, was kept open by the strong current, that this stream-hole must be regarded as one of the most unusual on account of its breadth, and that it is not at all decided if this strong current did not continue past Cape Jefferson, on which he stood, it appears probable that such a stream could continue its thawing activity far past this point; and even if it were correct that there had not been ice 36 miles out before this channel-opening, there is, however, no reason to seek such distant causes as those which the author has assigned in order to explain this phenomenon in another manner. Should there really be an open Polar basin in the summer, or at certain other periods, there is at all events no reason to suppose that this Open Sea had been reached by this expedition.

In conclusion, let me touch on the coasts discovered on this expedition, as represented on the chart at the beginning of the first volume. They who know how deceptive it is to look at the configuration of such high mountains at a distance from the sea, how all melts together, islands are taken for continents, promontories for islands, and deep spacious fjords and sounds quite disappear, will certainly agree with me in admiring the boldness with which the opposite coast, from Cape John Barrow to Mount Parry, an extent of more than two degrees of latitude, which they approached at the very nearest, at a distance of 25 to 40 miles, is found marked out on the said map as a clearly defined connecting shaded line, making only a little curve towards the east, in order to limit the Open Polar Sea, and, as if to receive the Gulf-stream, said to flow from Nova Zembla, and lead it down through Smith Strait to Baffin Bay. The heights of the mountains, according to the guessed distances, are on the other hand just as remarkable as determining the distances without knowing the heights of the mountains. The farthest mountain-top that Morton saw—"*the most remote northern land known upon our globe*"—has been put at 2500 to 3000 feet, and 100 miles from Morton's last station. Notwithstanding this great distance, Morton saw however that the top was bare, and that it was striped vertically with projecting ledges. Beyond this *ultima Thule*, about 60 to 80 miles from Morton's farthest station, and as it seems partly behind the Cape which stopped his view, is indicated "*open sea*," Had Morton only passed round his cape he would possibly have seen fresh capes shooting forth incessantly until he reached Mount Parry, which might have been thus connected by a neck of land with Greenland, and again on the other side large bays and sounds might have opened themselves on the American side and broken off the line now so nicely laid down on his map.

I have thus exhausted the most important points respecting these discoveries, which are represented as the crowning glories of the expedition. These Polar expeditions were dispatched for the discovery of the North-West Passage and of the remains of the Franklin Expedition, and both these problems have been solved by British enterprise. So far as they fall short of the finding the remains of Franklin or of the North-West Passage, they do not promise any advantages that can in any way answer to the means and efforts they demand.

Dr. Kane has undeniably gone beyond what he promised in his preface, namely, to give a simple narrative of the adventures of his party ; and he has hereby, in my humble opinion, injured more than benefited his work ; and the numerous really interesting and remarkable elucidations concerning the nature of North Greenland, obtained by immense labour and rare efforts, are thereby in a manner cast in the shade. Every one who interests himself for the Arctic regions will, in Kane's work, find valuable contributions to their description. Let me, among others, especially point out the description of the mode of life of the inhabitants of those northern regions; the remarkable abundance of walruses, bears, and other animal life; the observations on the growth of plants, and on the temperature, as well as those respecting the formations of ice on sea and land, &c. &c.[1]

---

[1] *See* 'Proceedings R. G. S.,' vol. ii. pp. 195 and 359, also vol. iii.—ED.

IV

THE ARCTIC CURRENT AROUND GREENLAND.

By Admiral E. Irminger, of the Danish Navy.[1]

Several hydrographers[2] assert that a current from the ocean around Spitzbergen continues its course along the E. coast of Greenland, and thence in a nearly straight line towards the banks of Newfoundland. In this opinion I do not agree, and give my reasons as follows.

Considerable quantities of ice are annually brought with the current from the ocean around Spitzbergen to the s. and s.w. along the E. coast of Greenland,[3] around Cape Farewell, and into Davis Strait.

These enormous masses of ice are frequently drifted so close to the southern part of the coast of Greenland that navigation through it is impossible. Experience has taught the captains who every year navigate between Copenhagen and the Greenland colonies (which all are situated on the w. side of Greenland) that, on going

---

[1] From the 'Journal of the Royal Geographical Society,' vol. xxvi.

[2] Korhallet. Berghaus, and others.

[3] See Grah, Scoresby, &c., as well as the 'Accounts of the Whalers in the year 1777. by Larens Hansen, Director of the School at Ribe,' in Denmark. These last-mentioned accounts of ten whalers, with their captains, and printed letters from several of these captains to the above-mentioned L. Hansen, give a striking proof of the current and its rapidity from the ocean around Spitzbergen to the s.w. along the E. coast of Greenland. The said ten vessels were enclosed in the ice in June 1777, in about 76° lat. N., between Spitzbergen and Jan-Mayen island, and were carried, constantly enclosed by the ice, in a south-westerly direction, between Iceland and Greenland, very often in sight of the Greenland coast. By degrees all the vessels were lost, being crushed by the ice ; the last vessel on the 11th of October, in 61° lat. N , in sight of Greenland. Of the crews of these vessels, which consisted of about 450 men, only 116 (whose names I have before me) were so fortunate as to save their lives, and get ashore from the ice in the month of October and beginning of November, on the coast around Cape Farewell. By calculating the distance between Cape Farewell and the place where the vessels were enclosed in the ice between Spitzbergen and Jan-Mayen, it gives a distance of about 1100 nautic miles, and the time the ice occupied in drifting from the above-mentioned place to Cape Farewell being about four months, the rapidity of this current has a mean of at least between 11 and 12 nautic miles per 24 hours.

H

to these colonies, in order to avoid being beset in the ice, they are obliged to pass a couple of degrees to the southward of Cape Farewell, as well as, after having crossed the meridian of this cape, generally not to steer much to the northward before reaching long. 50° or 52° w. of Greenwich, and sometimes even more westerly. The amount of westing is dependent on the wind, weather, or ice ; and by proceeding thus an open sea is reached, either quite free from ice or else with it much more diffused than near the coast, where the ships would be liable to be caught in the drifting masses.

A similar caution is exercised on the homeward passage from the colonies, the course being in the first place off the land, and then in a more southerly direction in order to reach the open sea free from the dangerous ice.

To be enabled to give an idea about the limits of the ice in these regions, I examined a set of logbooks which were kindly given me for perusal from the directors for the " Royal Greenland Commerce," viz., two logbooks for each of the last five years, which gives two outward and two homeward voyages to the colonies every year, consequently in all twenty voyages, which I found sufficient without extending these researches to too great a length.

There are unquestionably great changes in the limits of the ice in different seasons ; but still it is probable that the result of these five years' observations will not be far from the mean.

From these logbooks I noted at what latitude the meridian of Cape Farewell had been crossed on the passage to the colonies, and at what place the first ice was seen, and on what latitude the meridian of Cape Farewell was crossed on the homeward passage, and where the last ice was seen.

In the ensuing Table these positions are inserted, and, to make the subject still clearer, the places where the first and the last ice was seen are marked in the subjoined Plan.

By examining this Table it will be seen that the meridian of Cape Farewell is crossed on the outward passage in a mean lat. of 57° 46', and on the homeward passage in 58° 2' N., which gives 123 m. and 107 m. s. of Cape Farewell [1] respectively as the points where the ocean, according to the logbooks, *has been quite clear of ice,* and where, under ordinary circumstances, a safe passage can be made to avoid the ice, which is usually carried round the coast of Cape Farewell by the current coming from the ocean around Spitzbergen.

---

[1] According to the observations of Captain Graah, Cape Farewell is situated in 59° 49' lat. N., and 43° 54' w. of Greenwich.

Years	Vessel	Commander	Meridian of Cape Farewell passed	Outward Voyage to Greenland				Homeward Voyage from Greenland				
				Date	First Ice seen Lat. N.	Long. W.	Date	Meridian of Cape Farewell passed	Date	Last Ice seen Lat. N.	Long. W.	Date
1849	Brigantine Neptune	Knudsen	58 0 N.	July 26	58 10	47 19	July 28	58 5 N.	Aug. 31	58 21	44 35	Aug. 31
—	Barque Julianehaab	Bohn	59 23 ..	May 10	{Much ice at 20 m. N. of C. Farewell}		May 10	57 45 ..	Aug. 31	59 50	46 30	Aug. 29
1850	Brig Hvalfisken	E. Humble	58 7 ..	June 6	59 59	50 0	June 12	58 16 ..	Sept. 21	60 0	47 20	Sept. 21
—	Brig Egedesminde	Moberg	57 10 ..	May 28	60 30	52 10	June 8	57 50 ..	Sept. 23	61 31	53 53	Sept. 23
1851	Brig Lactnde	Faltings	58 7 ..	April 26	58 26	50 5	May 2	58 35 ..	Oct. 2	58 30	39 30	Oct. 4
—	Brig Mariane	Deger	57 30 ..	July 1	57 55	47 52	July 3	58 9 ..	Sept. 12	59 44	46 46	Sept. 11
1852	Brig Baldur	F. Ocken	57 43 ..	April 7	60 5	52 30	April 18	58 9 ..	June 5	60 15	57 14	May 27
—	Ditto, 2nd voyage	F. Ocken	55 53 ..	Aug. 10	Anchored at Godhavn, Aug. 19, without seeing any ice on the voyage			56 36 ..	Oct. 13	60 45	53 5	Oct. 3
1853	Brig Baldur	F. Ocken	57 56 ..	May 1	63 50	56 0	May 10	57 54 ..	June 10	61 7	61 13	June 5
—	Brig Pera	Humble	57 45 ..	June 23	59 41	52 25	June 26	58 53 ..	Sept. 17	60 40	46 45	Sept. 16

On the voyages from the colonies to Copenhagen the course pursued has been somewhat nearer Cape Farewell (16 m.), the cause of which is—1, that the captains, in coming from Davis Strait, have a better knowledge of the situation of the ice, and its distance from the land, than they can have on going up to Greenland in coming from the Atlantic Ocean, where no ice is to be seen; and 2, because the home passages are made in a season in which the ice generally is not quite so abundant as in spring, the season for the voyages to the colonies.

The subjoined Table shows that the brig *Lucinde* fell in with ice farthest to the E. (4th October, 1851, in 58° 30′ N., and 39° 30′ W. of Greenwich), which gives 79 nautic m. s., and about 135 nautic m. E. of Cape Farewell. This ice consisted only of a single isolated floe of very small extent; and it is very rare to meet ice in this latitude so far to the eastward.[1]

On the passage from Julianshaab to this place very little ice had been in sight.

On these voyages the first and the last seen ice generally consisted *of isolated icebergs or floes*, which no doubt formed the *very extremity of the ice* which was coming from the N.E. around Cape Farewell, and going into Davis Strait. Consequently the great and more accumulated masses of ice carried by the current from the ocean around Spitzbergen (whereby this current is really indicated) are *between these above-named outer limits and the coast of Greenland.*

The southerly and south-westerly coasts of Greenland are most exposed to be blocked up with these ice-drifts *in spring;* whilst, on the contrary, they are pretty clear of ice from September to January; but in the end of this month the ice generally begins to come again in great abundance, passing around Cape Farewell. (*Captain Graah*, p. 59.)

Still further to demonstrate the existence of this ice-drift, I may mention the following extract from the logbook of the schooner *Activ*, Captain J. Andersen. This vessel belongs to the colony of Julianehaab, and is used as a transport in this district :—

7th of April, 1851, the *Activ* left Julianshaab, bound to the different establishments on the coast between Julianshaab and Cape Farewell. The same day the captain was forced by the ice to take

---

[1] On the voyage to Greenland in 1828, Captain Graah fell in with the first ice in 58° 52′ lat. N., and 41° 25′ W. Greenwich, which is only 57′ s., and about 77 nautic miles to the eastward of Cape Farewell ; and he says, "Since 1817, I do not know that the ice has been seen so far to the eastward of the Cape."— 'Narrative of an Expedition to the East Coast of Greenland, by Capt. W. A. Graah, Royal Danish Navy,' p. 21, Eng. Transl

refuge in a harbour. Frequent snow-storms and frost. On account
of icebergs and great masses of floe-ice enclosing the coast, it was
impossible to proceed on the voyage before the 23rd, when the ice
was found to be more open; but after a few hours' sailing the
ice again obliged the captain to put into a harbour. Closed in by
the ice until the 27th. The ice was now open, and the voyage proceeded until the 1st of May, when the ice compelled him to go into
a harbour.

In this month violent storms, snow, and frost. From the most
elevated points ashore very often no extent of sea visible; now and
then the ice open, but not sufficiently so for proceeding on the
voyage.

At last, on the 6th of June, in the morning, the voyage was continued; but the same evening the ice enclosed the coast, and the
schooner was brought into "Blieschullet," a port in the neighbourhood of Cape Farewell.

The following day the voyage was pursued through the openings
between the ice; and the 18th of June the schooner arrived again
at Julianshaab.

Whilst the masses of ice, as above mentioned, enclosed the coast
between Julianshaab and Cape Farewell, the brig *Lucinde* crossed
the meridian of Cape Farewell on the 26th of April, in lat.
58° 3' N. (101 nautic m. from shore), and no ice was seen from the
brig before the 2nd of May, in lat. 58° 26' N., and 50° 9' W. of
Greenwich.

Further, Captain Knudsen, commanding the *Neptune* bound from
Copenhagen to Julianshaab, was obliged on account of falling in
with much ice, to put into the harbour of Frederikshaab on the
8th of May, 1852, and was not able to continue his voyage to
Julianshaab before the middle of June, because a continuous ice-
drift (icebergs as well as very extensive fields) was rapidly carried
along the coast to the *northward*.

Captain Knudsen mentions, that during the whole time he was
closed in at Frederikshaab he did not a single day discover any
clear water even from the elevated points ashore, from which he
could see about 28 nautic miles seaward.

Whilst the *Neptune* was enclosed by the ice at Frederikshaab
the brig *Baldur*, on the home passage from Greenland to Copenhagen (see the foregoing Table), crossed the meridian of Cape
Farewell the 9th of June in lat. 58° 9' N. (100 m. from shore) in
clear water, and no ice in sight.

From the above it is evident that the current from the ocean
around Spitzbergen, running along the E. coast of Greenland past

Cape Farewell, continues its course along the western coast of Greenland to the N., and transports in this manner the masses of ice from the ocean around Spitzbergen into Davis Strait.

If the current existed, which the before-named writers state to run in a direct line from East Greenland to the banks of Newfoundland, then the ice would likewise be carried with that current from East Greenland: if it were a submarine current, the deeply-immersed icebergs would be transported by it; if it were only a surface-current, the immense extent of field-ice would indicate its course,[1] and vessels would consequently cross these ice-drifts at whatever distance they passed to the southward of Cape Farewell. *But this is not the case:* experience has taught that vessels coming from the eastward, steering their course about 2° (120 nautic m.) to the southward of Cape Farewell, seldom or ever fall in with ice before they have rounded Cape Farewell and got into Davis Strait, *which is a certain proof that there does not exist even a branch of the Arctic current which runs directly from East Greenland towards the banks of Newfoundland.*

Along the E. coast, and around the southern and south-western coast of Greenland, the district of Julianshaab, there is generally a much greater accumulation of ice[2] than is the case more northerly, on the W. coast, or farther out in Davis Strait, where the ice generally is found more spread, and consequently it frequently happens that vessels bound to Julianshaab from Copenhagen are obliged first to put into some harbour more to the northward, and wait there until the ice is so much dispersed round the S. coast that they can continue their voyage to Julianshaab.

In the warmer season, when the ice and snow melt ashore, the waters from the different fiords or inlets move towards the sea, and drive the ice off the coast in such a manner that there is clear water close in shore, through which vessels may be navigated. However, continuing gales, according to their direction to or from shore, have an influence on the situation of the ice.

Another proof that the current from East Greenland does not

---

[1] An observation which it is interesting to mention here, and which gives a proof of the very little difference between the temperature of the surface and that at some depth, is mentioned in the Voyage of Captain Graah, p. 21. He says, "The 5th of May, 1828, in lat. 57° 35' N., and 36° 36' W., Gr., the temperature of the surface was found 6°·3 (46°·2 Fahr.), and at a depth of 660 feet 5°·5 + R. (44°·5 Fahr.)." This proves that there is no cold submarine current in the place alluded to in the S.E. of Cape Farewell. A still more conclusive experiment is recorded by Sir Edward Parry in the account of his first voyage, June 13, 1819: in lat. 57° 51' N., long. 41° 5', with a very slight southerly current, the surface temperature was 40½° Fahr.; and at 235 fathoms 39°, a difference of only 1½°.—ED.

[2] Captain Graah, pp. 10, 12, 22, 57, &c., English translation.

run in a straight line towards the banks of Newfoundland, is also derived from the observations of the temperature of the surface made on many voyages to and from Greenland. I have noted the observations of two voyages in the subjoined map;[1] one voyage by Captain Graah to Greenland, in May, 1828; and the other by Captain Holböll, from Greenland to Copenhagen, in September, 1844.

Captain Graah, who during his researches in Greenland, passed two summers and one winter on its eastern coast, between Cape Farewell and 65½° lat. N., says that he never found the temperature of the sea here higher than $0°·9 + R.$ ($34°$ Fahr.)[2]

Supposing that the Arctic current from East Greenland pursued its course in a straight line towards the banks of Newfoundland, it would be crossed, on the voyages from Copenhagen to the Danish colonies in Greenland, between $38°$ and $45°$ w. Gr., and so high a temperature in the surface of the ocean as from $4°$ to $6°$ R. ($41°$ to $45°·5$ Fahr.), as is found on this route and marked in the plan would, according to my opinion, be impossible, only $1°$ or $2°$ to the southward of the parallel of Cape Farewell; as it is a well-known fact that the principal ocean currents maintain their temperatures through very considerable distances of their courses.

This comparatively high temperature of the surface of the ocean so near to the limits of that current which carries enormous masses of ice from the ocean near Spitzbergen round Cape Farewell, warrants my opinion that the waters of the Atlantic Ocean move in a N.-westerly or northerly direction, towards the eastern and southern coasts of Greenland,[3] and that this in-draught towards the land is undoubtedly the cause of the ice being so closely pressed on to these parts of the coast as it is so frequently on the S. coast, and almost constantly on the E. coast, rendering the eastern coast entirely inaccessible from seaward.[4]

[1] This map is not found in the Society's 'Journal.'

[2] Graah says, "The temperature of the sea was frequently observed during the whole voyage, and was always found between 28° and 34° Fahrenheit.

[3] Graah says in his Narrative (p. 23, English translation), — "In the mouth of Davis Strait I found the temperature of the surface of the ocean from 4° to 3°·1 R. (41° to 39° Fahr.), though we were in the proximity of the ice. From this I concluded that a current from the *South* predominated here, because I never before in the vicinity of ice had found the temperature of the water exceeding 1°·8 R. (36° Fahr.), and this conclusion was confirmed when, coming to the northward of the ice, I found the temperature of the water 1°·1 + R. (34°·5 Fahr.)"

[4] Besides the evidence afforded by the ice-drifts and the temperature of the water, as cited by the author, conclusive proof of a northerly set is found in the driftwood which has been so frequently met with around Cape Farewell and off the w. coast of Greenland. A few examples will suffice. A plank of mahogany was drifted to Disco, and formed into a table for the Danish governor at Holsteinborg ('Quarterly Review,' No. xxxvi.). Admiral Löwenörn picked up a worm-

The logbooks which I have examined afford no positive information as to the direction and force of the current under consideration—a circumstance which must be attributed to the frequency of fogs and gales of wind, which prevent correct observations being made.[1]

From the foregoing it seems to me to be demonstrated that the current from the ocean around Spitzbergen, which carries so considerable masses of ice, after it has passed along the E. coast of Greenland, turns westward and northward round Cape Farewell, *without detaching any branch* to the south-westward, directly towards the banks of Newfoundland.

This current afterwards runs northward along the s.w. coast of Greenland until about lat. 64° N., and at times even up to Holsteinborg, which is in about 67° N.

This current undoubtedly afterwards, by turning to the westward, unites with the current coming from Baffin and Hudson Bays, running to the southward on the western side of Davis Strait along the coast of Labrador, and thus increases that enormous quantity of ice which is brought towards the s. to Newfoundland and further down in the Atlantic Ocean, frequently disturbing and endangering the navigation between Europe and Northern America.

---

eaten mahogany log off the s.e. coast of Greenland. These in all probability were transported from the s.w. by the Gulf-stream. Captain Sir Edward Parry, in his second voyage, September 24th, 1823, picked up a piece of yellow pine quite sound, in lat. 60° 30′, long. 61° 30′ w.; and on his third voyage seven pieces of driftwood were found in the vicinity of Cape Farewell. Again, Captain Sir John Ross found much driftwood around Cape Farewell; and Captain Sir George Back saw in lat. 56° 50′, long. 36° 30′, a tree with the roots and bark on. These instances might be multiplied, but their character indicates a southern origin.—ED.

[1] Sir John Ross, in his first voyage, May 23, found the current to run 6 m. per day to the w.n.w. in lat. 57° 2′ and long. 43° 21′ w. (or about 168 m. s. of Cape Farewell), and n.w. when 140 m. s. by w. of the Cape.

Sir Edward Parry, on June 19, 1819, when 130 m. due w. of Cape Farewell, found its direction and velocity to be s. 50° w. 6 m. per diem.—ED.

V.

NOTES ON THE STATE OF THE ICE

And on the INDICATIONS of OPEN WATER, &c., from BEHRING to
BELLOT STRAITS, along the COASTS of ARCTIC AMERICA and
SIBERIA, including the ACCOUNTS of ANJOU and WRANGELL.

INTRODUCTION.

IN undertaking that part of the Arctic manual which has been
assigned to me by the Council of the Royal Geographical Society, I
feel that I am to a great extent occupying a portion of the Arctic
Sea which can be of little importance to the present expedition.
Yet there is no doubt that the influence which the Pacific Ocean
exercises on the motion of the ice should be considered in the present
attempt to reach the Pole. I have, therefore, first endeavoured to
give an account of the different voyages by which the exploration
of this area has been carried forward, and then to summarise the
result. I then purpose to give a short account of the descent of
the Mackenzie, the Coppermine, and the Back Rivers, together
with the exploration of the coast in boats and birch-bark canoes
and the voyages of the *Investigator* and *Enterprise.* These, it is to
be hoped, will enable the reader to form a correct judgment respecting the state and the movement of the ice in the Polar Sea from
New Siberia to Bellot Strait.

In 1725 the Russian Government dispatched an expedition
through Siberia to the sea of Ochotsk, which they occupied two
years in reaching. They there built and launched the vessels,
which, proceeding to the north, discovered St. Lawrence Island,
and eventually reached the latitude of 67° 18′ N. How the Commander, in attempting during a second expedition to carry his explorations over to the American Continent, and how, by persevering
to his uttermost, he came by his death, cannot find a place here ;
but we who are able to appreciate the difficulties he had to encounter,
glory to think that the title of Behring is handed down to posterity
by the name so justly given to the sea and the strait which separate
the continents of Asia and America.

Beginning with a short *résumé* of Cook's voyage in this vicinity, I then take up the Russian explorations, which, from their interesting character, bearing as they do so directly on the important question of the Polynia or open sea, I have found great difficulty in comprising within a small compass. I then give a short account of the valuable and instructive voyage of the *Blossom*, under Captain Beechy. A general historical account of the expeditions in search of Sir J. Franklin, by way of Behring's Strait, follows. These I have endeavoured to render as short as possible, with the exception of the part taken by the *Enterprise*, which has been given more at large, under the impression that it may be the means of preserving information that might (as has been the case in other voyages) pass away without a knowledge of the value that notes daily made have upon future researches.

Some observations have been collected from Mr. Whymper's interesting narrative of his adventures in this sea, which I have little doubt will prove useful.

Some extracts from the correspondence of the American whale-fishers have been added, which will elucidate the change in the position of the ice in different years.

In collating the general information with which I sum up this portion of the work, I have had recourse to official documents as well as publications by private individuals; but I am especially indebted to two officers who have spent five seasons in this neighbourhood. One I regret to say is no more; but the valuable information collected by Dr. Simpson will leave a regret that he was not spared to carry further the extent of his ability: the other is Captain Hull, at present Assistant-Hydrographer, who, after three summers spent in Behring Sea in the *Herald*, returned with Commander Maguire and passed two winters in the *Plover* at Point Barrow.

1.—BEHRING STRAITS.

In the instructions issued by the Admiralty to Captain Cook, when proceeding on his third voyage in July, 1776, the following paragraphs appear :—" Upon your arrival upon the coast of New Albion, you are to put into the first convenient port to recruit your wood and water and procure refreshments, and then to proceed northward along the coast as far as latitude 65°, or farther, if you are not obstructed by lands or ice. . . . . When you get that length, you are very carefully to search for and to explore such rivers or inlets as may appear to be of considerable extent, and pointing towards Hudson or Baffin Bays."

" In case you shall be satisfied that there is no passage through to the above-mentioned bays sufficient for the purpose of navigation, you are at the proper season of the year to repair to the port of St. Peter and St. Paul, in Kamschatka, or wherever else you shall judge more proper, in order to refresh your people and pass the winter; and in the spring of the ensuing year, 1778, to proceed from thence to the northward, as far as in your prudence you may think proper, in search of a north-east or north-west passage from the Pacific into the Atlantic Ocean or North Sea."

In accordance with these instructions, the *Resolution* and *Discovery* sailed from Oonalashka on July 2nd, 1778, and upon the 9th of August reached Cape Prince of Wales, which was then pronounced by Captain Cook (as it eventually has been proven to be) the western extremity of all America hitherto known. The ships then proceeded to the north, and reached the edge of the ice in lat. 70° 41' on August 17th, naming the furthest point on the American shore Icy Cape. They then visited the Asiatic side of the Straits and discovered Cape North. The ships bore up on August 29th, and after making farther explorations on the American shore south of Cape Prince of Wales, returned to Oonolashka on October 2nd. The following are Captain Cook's remarks after visiting the pack in a boat:—" I found it consisting of loose pieces of various extent, and so close together that I could hardly enter the outer edge with the boat, and it was as impossible for the ships to enter it as if it had been so many rocks. I took particular notice that it was all pure transparent ice, except the upper surface, which was a little porous. It appeared to be entirely composed of frozen snow, and to have been all formed at sea. . . . . The pieces of ice that formed the outer edge of the field were from 40 to 50 yards in extent to 4 or 5, and I judged the latter pieces reached 30 feet or more under the surface of the water."

The following year the ships left the harbour of St. Peter and St. Paul on June 13th. At noon on the 6th of July, in lat. 67°, large masses of ice were fallen in with; on the 8th, in lat. 69° 21', they were close to what appeared from the deck solid ice. On the 18th of July they reached 70° 26' N., and were close to a firm united field of ice. By stretching over towards the American Continent they reached 70° 33', the farthest northern point attained this season, when, finding it impracticable to get any farther, Captain Clerke determined to expend the remainder of the season in endeavouring to find an opening on the Asiatic coast. In doing so the *Discovery* was beset in the ice in lat. 69°, when she was drifted to the north-east at the rate of half a mile per hour, and was so much damaged, that after in vain attempting to get to the north between the ice and the land, Captain Clerke, on July 23rd, determined to proceed to the southward, and reached the harbour of St. Peter and St. Paul on August 24th.

Russian Explorations East of the River Kolyma.—In the year 1762. the River Kolyma was descended by Schalarov, and the coast explored as far as Cape Chelagskoi. In the same year Sergeant-Andrejew discovered the Bear Islands.

Hedenstrom, who explored the coast of Siberia between the years 1808 and 1811, makes the following remarks:—"The shores of the Polar Ocean from the Lena to Behring Strait are, for the most part, low and flat, rising but little above the level of the sea, that in winter it is difficult to tell where the land terminates. A few wersts, however, inland a line of high ground runs parallel with the present coast, and formerly, no doubt, constituted the boundary of the ocean. This belief is strengthened by the quantity of decayed wood found on the upper level, and also by the shoals which run out far to sea, and are probably destined at some future period to become dry land. On these shoals during the winter lofty hummocks of ice fix themselves, forming a kind of bulwark along the coast, and often remaining there during the whole summer without melting. The nearer the Arctic shore is approached, the more scanty and diminutive the trees become. Beyond 70° neither trees nor shrubs are met with.

In the year 1786 Billings built two boats—one 45 feet and the other 28 feet long—at Jássaschna on the Kolyma, and left the entrance of the river in them on June 27th. The ice frequently compelled them to run into bays, and take shelter under headlands. On July 1st they attempted to sail to the north, and were not able to get more than 20 miles from the shore when " the whole sea, as far as the eye could reach, being covered with immense masses of ice, on which the waves broke with tremendous violence," they were obliged to turn back. Constant ice and frequent fogs impeded them so much, that they did not pass the Great Baranov Rock before July 19th. Eleven miles further they came to ice hummocks aground in 16 fathoms water, when, it being impossible to go further, they returned to the mouth of the Kolyma on July 26th.

In 1788 Billings sailed from Avatska and put into St. Lawrence Bay, where the Tchutskis told him the sea was covered with such quantities of ice that its navigation was impracticable. He therefore relinquished his plan of sailing to Cape Chelagskoi, and determined to proceed by land. Returning in the ship to Metchigme Bay, he commenced his journey on reindeer sledges to Koliutchen Bay. Sending his companion Gilew in a Tchutski baidar with orders to survey the coast from East Cape to Koliutchen Island, Gilew followed the coast to East Cape, where the ice was pressed so closely on the shore that he was compelled to drag the baidar across a narrow neck of land. He then followed the coast until within 90 miles of Koliutchen Island, when the Tchutski refused to go any further. Fortunately he fell in with a tribe of reindeer Tchutski, who conducted him to Koliutchen Bay, where he met with Captain Billings. After surveying the shores of this bay, Billings proceeded with the Tchutski to the first Russian settlement on the Aniui (a tributary of the Kolyma), where he arrived on February 17th. Speaking of the Tchutski land as barren in the extreme, he says, " before July there is no symptoms of summer, and on the 20th of August the winter sets in."

Baron Wrangell reached the mouth of the Kolyma on February 21st, 1821, temperature 26°, loading of each sleigh 1000 lbs. Leaving on the 22nd, the Great Baranika was reached on the 27th, where great quantities of drift-wood were found. At Chelagskoi Ness on March 5th ice hummocks were 90 feet high. They then proceeded 40 miles east of Cape Chelagskoi, and returned to Niznei Kolymsk on March 14th, having been absent twenty-two days, and travelled over 650 miles.

With eight men, besides dog-drivers, 240 dogs, and twenty-two sledges, carrying thirty days' provisions, he left the mouth of the Kolyma on March 26th, temperature + 21° F. At a mile from the shore they came to a chain of ice hummocks, and a wide fissure in the ice. After three hours' labour they got through the hummocks, and came upon an extensive plain of ice, broken only by a few scattered masses ; at noon on the 28th, in lat. 69° 58' N., numerous traces of foxes going in the same direction as ourselves. March 29th, temperature + 14°, at 4 P.M., reached the Bear Islands: an appearance of open water to the N.N.W. There was a much greater quantity of drift-wood on the north than the south side of the islands. March 31st, wind north-east, temperature + 7° A.M. + 14 P.M., came upon sharp grains of seasalt ; snow more soft and damp ; fog so moist as to wet our clothing. Camped under a wall of ice 30 feet high. Thickness of ice 3½ feet ; lat. 70° 53' N.

April 1st.—Thermometer + 23 A.M. + 7 P.M. After pursuing a N. by E. course for 14 miles tracks of foxes were seen, and the ice

hummocks contained earth and sand. Travelling difficult; seven
hours in accomplishing 19 miles.

April 2nd.—Wind north-west; snow; temperature + 18; course
N. by W. Many hummocks; ice 1 foot thick; three seals seen;
depth of water 12 fathoms; green mud. Camped in lat. 71° 31' N.

April 3rd.—Thermometer + 16°, fox-tracks from W.S.W. to E.N.E.
Ice only 5 inches thick, and very rotten. Felt the undulatory
motion under the ice; lat. 71° 37'.

April 4th.—A gale from the north; temperature + 16°. Proceeded northerly with two sleighs; reached lat. 71° 43'; ice so rotten
that we were compelled to return. The hummocks are sometimes
80 feet in height.

April 5th.—Wind S.S.E. ; temperature + 9' A.M. + 7 P.M. ; in
lat. 70° 30'.

April 6th.—Wind south-east; temperature + 18 A.M., and − 2 P.M.;
ice agitated, and in the north-east loud noise of ice crushing
together; lat. 71° 15'.

April 7th.—Wind east ; temperature + 5 A.M. − 6 P.M. ; in
lat. 70° 56'. April 8th.—Wind south ; temperature 0. Came upon
a wide fissure; ferried across on a floating block of ice. Current
half a knot in an E.S.E. direction; depth of water 12½ fathoms; ice
violently agitated, opening in various directions; lat. 70° 46'. On
April 13th reached lat. 71° 4', where eight dogs fell through the
ice into the water. Returned to Four Pillar Island on the 18th, and
to Niznei Kolymsk on the 28th, having been absent thirty-six days,
and have travelled 700 miles with the same dogs.

In 1822 Baron Wrangell left Niznei Kolymsk on the 10th of
March, with five travelling and nineteen provision sledges carrying
provision for forty days. The Baranov Rock was reached on the
14th, and on March 23rd the thermometer rose to + 35° in lat.
70° 42' and 1° 50' E. of Baranov Rock. On April 9th, in lat. 71° 51'
and 3° 20' E. of Baranov Rock several fissures were met with, in
which a depth of 14½ fathoms green mud was found. Here the
ice hummocks prevented the heavy-laden sleighs proceeding further ;
a light sleigh proceeded 6 miles to the north, when all progress was
stopped by the complete breaking up of the ice and a close approach
to the open sea. On the 19th, in lat. 71° 18' and 4° 36' E. of
Baranov, a depth of 21 fathoms was found, with a rather strong
current running to the E.S.E. On the 22nd Cape Chelagskoi was
sighted, and they returned, on May 5th, to Niznei Kolymsk, having
been absent 57 days and travelled over 782 miles.

In 1823 Baron Wrangell left Baranika on March 5th and reached

Cape Chelagskoi on the 8th. Here the first Tchutski were met, who told them the native name for the cape was Erri, and the name of Cape North Ir-kai pi, and " that between these capes, from the top of some cliffs near the mouth of the river, one might, in a clear summer's day, descry snow-covered mountains at a great distance to the north, but that in winter it was impossible to see so far." (This is the first notice of the land afterwards discovered by Capt. Kellett, in the *Herald*.) He had been told by his father that a Tchutski had once gone there in a 'baidar,' and he thought the northern land was inhabited.

On the 10th the journey was continued; several large heaps of whalebone were seen, but very little drift-wood; on arrival at Schalarov Island (which is called Amgaoton by the natives) an attempt was made to go to the north, but on the 21st, in lat. 70° 20′ N. and long. 174° 13′ E., they were compelled to return, " the hummocks now becoming absolutely and entirely impassable," and a break in the ice was fallen in with, extending east and west farther than the eye could reach and 150 fathoms across at its narrowest part. The current was running $1\frac{1}{2}$ knot to the eastward, depth of water $22\frac{1}{2}$ fathoms, in lat. 70° 51′, long. 175° 27′ E. " Fragments of ice of enormous size were thrown by the waves with awful violence against the edge of the ice-field." The coast to the eastward was then followed, and habitations of the Tchutski as well as driftwood found, " which, in all probability, is of American origin; eventually Cape North was reached on April 11th.

The shores of the Bay of Anadyr are inhabited by a people distinct from the Tchutski in figure, countenance, clothing, and language, called Onkilon or Sea People; there are traditions that two centuries ago the Onkilon occupied the whole coast from Cape Chelagskoi to Behring Strait. Whales are particularly abundant in the neighbourhood of Koliutchen Island.

On May 1st Cape Chelagskoi was reached, and Baron Wrangell returned to Niznoi Kolymsk, after an absence of 78 days, having accomplished a distance of 1327 miles, or 17 miles per diem.

M. Von Anjou took his departure from the north-west end of Kotelnoi (the west island of New Siberia), on April 5th; at a short distance from the shore it was found necessary to open a path through the ice hummocks by crowbars; at 20 miles from the island they had 15 fathoms mud; in lat. 76° 38′ N. they found 17 fathoms, and " the near vicinity of the open sea forbid further progress." They then crossed over to Fadejevskoi Island, from whence " dense vapours " are seen, indicating the vicinity of the open sea.

The expedition then crossed to the eastern island of the group, which they reached on the 18th, and saw " to the north the open sea with drift-ice." At Cape Raboi " the ice appeared unbroken."

In 1822 M. Von Anjou left Svatoi Ness on April 10th, reached Leakhow Island on the 12th, and Kotelnoi on the 18th ; following the east coast, the north extremity was rounded, from whence they attempted to go north, but were stopped by thin ice ; proceeding easterly along its edge, land was seen to the s.s.w., which proved to be a low island to which the name of Figurin was given.   On its shores " were drift-wood of larch, traces of bears and grouse, and old nests of geese were found."   They afterwards went 15 miles n.w. by n., across large hummocks. when their progress was again arrested by thin ice ; here they had 10 fathoms sand ; at last they came to open water, in which, " though the wind was westerly," the pieces of ice were drifting from east to west.   The sledge drivers were of opinion " that this current was the ebb tide, the regular six-hourly return of which they had noted."

They then went 60 miles in a north-easterly direction from Cape Kaimenoy, when the thinness of the ice stopped them, and they had a depth of 15 fathoms mud.

They reached Niznei Kolymsk on May 5th.

Baron Wrangell's Remarks.—The fur-hunters, who visit North Siberia and Kotelnoi Island every year and pass the summer there, have observed that the space between these islands and the continent is never completely frozen over before the last days in October. In the spring the coasts are quite free by the end of June. Winter hummocks are frequently 100 feet high. The great Polynia, or that part of the Polar Ocean which is always an open sea, is met with about 4 leagues north of New Siberia, and from thence, in a more or less direct line, to about the same distance off the continent between Cape Chelagskoi and Cape North. During the summer the current between Svatoi Ness and Koliutchen Island is from east to west, and in autumn from west to east. North-west winds prevail in the spring. The inhabitants of the north coast of Siberia generally believe that the land is gaining on the sea, and this belief is chiefly founded on the quantity of long withered drift-wood which is now to be met with on the tundras and in the valleys 20 miles from the present sea-line, and decidedly above its level.

The *Rurik*, under the command of Lieutenant Von Kotzebue, arrived off Behring Island on June 20th, 1816.   He landed on

St. Lawrence Island on the 27th, reached Cape Prince of Wales on the 30th, and discovered Kotzebue Sound on August 1st. After exploring it he returned to the south on the 19th. An account of this voyage, in three volumes, was published in London in 1821.

Captain Lutke, in command of the corvette *Le Seniavine*, visited Behring Strait in the years 1828 and 1829. To him we are indebted for some valuable charts of the coast of Asia between the 53° and 65° of lat. A narrative of this portion of the voyage will be found at Chap. XI. vol. ii., and from pages 17 to 56 in vol. iii. of the account of the voyage, which was published in Paris in 1835.

Voyage of the ' Blossom.'—The *Blossom* left the harbour of St. Peter and St. Paul on July 5th, 1826. On the 18th, when off St. Lawrence Island, a current was found to be setting to the north-east, three-quarters of a mile per hour. On the 20th, the Diomede Isles were reached, and Kotzebue Sound on the 22nd. Leaving that place on the 30th, the current off Point Hope was found to be superficial, *i.e.*, not extending 12 feet below the surface; but at the surface it attained a rate in a westerly direction of 3 miles per hour. In the evening it slacked to 1½ miles. On the 8th of August Cape Lisburn was reached, and the edge of the pack was fallen in with in lat. 71° 8′ N. on the 13th. The barge, under the charge of Mr. Elson was sent to the northward on the 17th, and the *Blossom* returned to Kotzebue Sound on the 28th. The barge arrived here on the 10th of September, having reached Point Barrow in 71° 23½′. In returning she was driven on shore by the pack; but the wind coming round to the S.S.E., she made her escape with much difficulty. The ice at Point Barrow was aground in 4 fathoms water, and was 14 feet above the level of the sea. On the 13th of October the thermometer fell to 27°, and the edge of the sound began to freeze. The *Blossom* proceeded to sea and reached the Aleutian Islands on the 22nd.

In the year 1828 the *Blossom* left St. Peter and St. Paul Harbour on July 18th. In crossing over to the American shore the temperature was found to be 21° higher. The ship arrived at Kotzebue Sound on the 5th of August. Leaving on the 16th, the pack-edge was reached on the 18th in lat. 70° 6′, or 24 miles south of its position on the 13th of the same month last year. On October 4th the thermometer fell to 25° in Kotzebue Sound. The ship left on the 6th, and reached the Aleutian Islands on the 14th. The following are Captain Beechey's remarks on the currents in Behring Strait:—

" It does not appear from our passages across the sea of Kamschatka that any great body of water flows towards Behring Strait. In one year the whole amount of current from Petropaulowski to St. Lawrence Island was s. 54°, w. 31 miles, and in the next N. 50°, w. 51 miles, and from Kotzebue Sound to Oonemak N. 79°, w. 79 miles.

" Approaching Behring Strait the first year with light southerly winds, the current ran north 16 miles per diem; and in the next, with strong south-west winds, north 5 miles; and with a strong north-east wind, N. 34°, w. 23 miles. By this it appears that near the strait, with southerly and easterly winds, there is a current to the northward, and with northerly and north-westerly winds there is none to the southward; consequently the preponderance is in favour of the former.

" To the north of Behring Strait the northerly current is more apparent. It was first detected off Schischmaroff Inlet; it increased to between 1 and 2 miles per hour off Cape Krusenstern, and arrived at its maximum 3 miles per hour off Point Hope: this was with the flood; with the ebb it ran w.s.w. half a mile per hour.

" Off Icy Cape the current appeared to be influenced by the winds. Near Point Barrow it ran at the rate of 3 miles per hour and upwards to the north-east, and did not subside immediately with the wind; but the current here must have been accelerated by the pack closing in on the beach.

" It is a curious fact that the margins of the ice between America and Asia, Europe and Greenland, lie as nearly as possible in the same direction, viz., south-west and north-east, and that the navigation on the west shores is impeded in a much lower latitude than the eastern.

" Near Icy Cape, south and west winds occasioned high tides; north and east, low ebbs. The tide rises about 2 feet 6 inches at F. and C., and the flood comes from the southward.

" From St. Lawrence Island there appears to be a current running to the north of about three-quarters of a mile per hour."

Expeditions in Search of Sir John Franklin.—H.M.S. *Herald,* Captain Kellett, arrived at Petropaulski on August 14th, and at Kotzebu Sound on the 1st of September, where she remained until the 29th.

The *Plover,* Commander Moore, left Honolulu on August 25th, reached the island of St. Lawrence on October 13th, went into harbour near Tchutski Ness in lat. 64° 20′ and long. 173° 15′ w. on the 25th, and was permanently frozen in on November 18th.

On June 13th a clear lane of water enabled the ship to put to sea, and she arrived at Chamisso Island, Kotzebue Sound, on July 14th.

The *Herald* joined the *Plover* in Kotzebue Sound on July 15th, and in company with the *Nancy Dawson* (Captain Shedden's yacht), proceeded to sea on the 18th. On arriving at Wainwright Inlet the boats were dispatched from the ships on the 25th instant, and Lieutenant Pullen, in command of them, reached Point Barrow on August 4th, in company with the *Nancy Dawson*. After seeing the boats fairly off on their route to the Mackenzie River, Mr. Shedden rejoined the *Herald* in Kotzebue Sound.

The *Herald* reached her northernmost lat. 72° 51′ N., in long. 163° 48′ W., on July 29th, and on August 17th an island was discovered, with a long range of high land beyond it. Captain Kellett thus describes the landing on the island:—"We reached the island, and found running on it a heavy sea. The First Lieutenant, Maguire, landed, having backed his boat in until he could get foothold. I followed his example; others were anxious to do the same, but the sea was so high I could not permit them. We hoisted the jack and took possession in the name of Her Majesty." "The extent we had to walk over was not more than 30 feet, from which we collected eight different species of plants." "The island is about 4½ miles in extent east and west and 2½ north and south, in the shape of a triangle, with the west end as the apex." "It is almost inaccessible on all sides, and a solid mass of granite." "Innumerable black and white divers here found a safe place to deposit their eggs: not a walrus or seal was seen either on the shore or the adjoining ice, and none of the small land birds."

Speaking of the land to the north Captain Kellett says:—"It becomes a nervous thing to report a discovery of land in these regions without actually landing on it, after the unfortunate mistake to the southward; but as far as a man can be certain, who has 130 pairs of eyes to assist him, and all agreeing, I am certain we have discovered an extensive land." The *Herald* returned to Kotzebue Sound on the 1st of September.

The *Plover*, Commander Moore, passed the winter of 1849–50 in Estcholtz Bay, Kotzebue Sound, and he thus describes the breaking up of the ice in the spring:—"As the bay cleared a little, giving the ice more play, the ship became much hampered, requiring the utmost vigilance to prevent her being pushed high upon the beach or overwhelmed with the pressure of floe upon floe, frequently depending upon her safety upon anchors and cables; and when the former, losing their hold in the ground, allowed her to drive, they still had the effect of keeping the stem to the pressure to which I conceive her safety was owing. On the 25th of June the ebb-tide of both morning and evening thus forced the ship on the ground. The floes, 3 to 4 feet in thickness, rising along the inclined plain of the cable, then splitting to the distance of several hundred feet ahead, were crushed beneath the stem or thrown outwards off the bows; then passing astern, piled in broken masses, 12 or 15 feet in height, along the shore of the bay." The *Plover* left Kotzebue Sound on July 17th and proceeded to the northward, leaving the ship off Wainwright Inlet on the 23rd, in two boats. Captain Moore reached Point Barrow on the 27th, and went round it as far as Dease Inlet, and returned to Grantly Harbour, Port Clarence, on the 30th, where she passed the third winter.

The *Herald* left Oahu on the 24th of May, 1850, and reached Kotzebue Sound on the 16th of July; and upon the 31st fell in with the *Investigator* off Cape Lisburne, and returned to Port Clarence on the 4th of September.

The *Investigator* left Oahu on July 4th, passed through the Aleutian chain of islands on the 20th, and reached Cape Lisburne on the 29th. Entered the ice in lat. 72° 1' N., and long. 155° 12' W., and at midnight on August 5th rounded Point Barrow in 73 fathoms water, 10 miles from the land. They got into open water on the American shore on the 7th, and landed at Port Drew on the 8th.

The *Enterprise* left the Sandwich Islands on June 30th, and passed through the Aleutian chain of islands on July 28th. East Cape was reached on August 12th, the total set of the currents in the intervening fifteen days being N. 49°, E. 127 miles.

Proceeding north, the following table will show the daily position of the ship at noon, and the current experienced in each 24 hours :—

Date.	Latitude.	Longitude.	Current.
August 13	68°56	166°19	N. 60° W. 19 miles.
,, 14	70·26	162·48	N. 21° E. 35 ,,
,, 15	70·39	160·6	S. 83° E. 7 ,,
,, 16	71·50	160·36	N. 3° W. 13 ,,
,, 17	72·44	159·03	N. 28° W. 8 ,,
,, 18	72·41	158·52	W. 9·4 ,,
., 19	72·29	158·49	N. 10 ,,

Passing Point Hope, Cape Lisburne, and Wainwright Inlet without seeing anything of the *Herald*, *Investigator*, or *Plover*, the *Enterprise* got up to the ice on the 16th, and pushing through some brash ice entered an open lane, trending north-east and south-west, 10 miles wide, up which she proceeded until she had gained a position north-east by north 100 miles from Point Barrow, and had 45 fathoms depth of water-mud. Here her progress was barred, and after searching in vain for any opening on the southern side, she was on the 21st within 30 miles of the land, without a prospect of reaching it. The ice hummocks here were frequently found to be 25 feet above the sea, and on one or two instances as much as 30 feet was seen; but this was the greatest height. Here no bottom with 60 fathoms was found, and the temperature of the sea rose to 40°. Seeing there was no hope of progress in this lane, the ship's head was turned to the southward, having traced the pack in a south-easterly direction for 145 miles from lat. 72° 45′ and long. 159° 5′ w., with no signs of opening to east or south-east. Further progress to the eastward was considered impracticable this season.

Returning to the south, the southern edge of the pack was found on August 27th to be 20 miles to the southward of where it was on the 16th. Rounding it a lane of open water was found trending e.n.e. and w.s.w., up which we proceeded, reaching at length lat. 73° 23′, in long. 164° 4′ w., where the pack-edge, both to the east and west, trended southerly, leaving no hope of further progress to the north or east.

Point Hope was reached on August 31st, and Port Clarence on the 2nd of September. On the 14th the *Enterprise* again proceeded to the north, and remained cruising between Cape Lisburne and Icy Cape until the 30th, when the thermometer fell to 18°, and the ice formed so rapidly as to interfere with navigation. We returned to Port Clarence on the 2nd, and visited Fort Michailowski, in Norton Sound, on the 16th of October, where Lieutenant Bernard, Mr. Adams, assistant-surgeon, and Thomas Cousins, A.B., were landed.

Leaving Hongkong on the 2nd of April, 1851, the *Enterprise* touched at the Bonin Islands on the 28th, passed through the Aleutian chain on May 24th, and fell in with the pack in lat. 62° and long. 179° e. on June 1st. Boring through the pack, Cape Behring was seen on the 11th. On the 18th three baidars came off to the ship from Cape Atchene, which was 8 miles distant. Cape Prince of Wales was sighted on the 25th, but owing to the quantity of ice in Behring Strait, Port Clarence was not reached until July 4th. The whole length of the passage was 41 days, of which 28 were passed in the pack. The position by account differs from that by observation n. 54° e. 264 miles, or $6\frac{1}{2}$ miles per day. Leaving Port Clarence on July 12th, Wainwright Inlet was reached on the 19th, and on the 20th the ship was beset in the pack off the Seahorse Islands, when the following currents were experienced :—

Date.	Time.	Set of Current.	Depth.
		per hour.	
July 20th.	1 A.M.	N. by E.	0·5
	Noon.	N.E. by E.	0·9
	4 P.M.	N.N.E.	0·4
	2 A.M.	N.E. by N.	0·6	17 fms.
July 21st.	8 A.M.	N.N.E.	0·5	18 fms.
	Noon.	N. tone.	0·6	15½ fms.
	10 P.M.	E.	0·4
July 22nd.	Noon.	E.	0·6
Lat. 71°·1'.	4 P.M.	E. by N.	0·4
Long. 158°·03'.	Midnight.	N.E. by E.	0·4	20 fms.
July 23rd.	4 A.M.	E. by N.	0·7
Lat. 71°·09'.	Noon.	N.E. by E.	0·3
Long. 157°·37'.	4 P.M.	N.E. by E. ½E.	0·5
	Midnight.	E.N.E.	0·6
	2 A.M.	E. by E.	0·6
July 24th.	Noon.	S.E.	0·5
Lat. 71°·12'.	4 P.M.	E.	0·5
Long. 156°·51'.	8 P.M.	N.N.E.	0·5
	Midnight.	N.	0·5
	2 A.M.	E.	0·8	16 fms.
	8 A.M.	N.N.W.	1·3	25 fms.
July 25th.	Noon.	N.E. by E.	1·8
Lat. 71°·27'.	2 P.M.	E.N.E.	1·3
Long. 156°·12'.	6 P.M.	N.E. by E.	2·1	14½ fms., sand.
	10 P.M.	E.N.E.	1·3	13 fms., sand.
	Midnight.	E.N.E.	1·5
July 26th.	10 A.M.	N.	1·3
Lat. 71°·31'.	Noon.	N.E. by E.	1·8	32 fms.
Long. 154°·50'.	2 P.M.	E.N.E.	1·3	24 fms.
Drifting past	8 P.M.	N.E. by E.	2·1	14½ fms.
Point Barrow	10 P.M.	N.N.E.	1·3	13 fms.
in the pack.	Midnight.	E.N.E.	1·5

(Left margin label for table: *Set of the current while rounding Point Barrow in the pack.*)

During the 27th, the ice opened at times, permitting us to make a little progress, nor could we perceive any current.

Throughout the whole of the 28th, we remained immovable, except for an hour about noon, when we managed to warp her through one or two holes, the depth of water varied from 13 to 11 fathoms mud. On the ice which was much broken up were many shells (*Nymphacca*), which at first were thought to have been brought there by birds: but, eventually, we came alongside a floe, on which were three large stones (greenstone, 30 to 50 lbs. weight): therefore the mass we were alongside of, and which was surrounded by ice as far as the eye could reach from the crow's nest, had been in contact with the shore this season: the nearest land was 10 miles distant. The ship remained alongside the stones the whole of the 29th. In the afternoon the current, which for the last twenty-four hours had been imperceptible, took a south-easterly trend. On the

30th, Point Barrow bore south-west by south, and there was open water to the south, not more than 3 or 4 miles distant, which we succeeded in warping into at 3.40 P.M., and, working between the ice and the shore, reached Port Tangent at noon on the 31st. The length of the passage from Cape Prince of Wales to Point Barrow is 13½ days, during which we were 4½ days beset in the pack. The total amount of current in the same time is N. 73° E. 172 miles, or 12·7 miles per day.

H.M.S. *Dædalus*, Captain Wellesley, after encountering much difficulty with the ice in the neighbourhood of St. Lawrence Island, reached Port Clarence on July 15th. The *Plover*, after receiving supplies, proceeded north, but, finding the pack very far south, returned to her winter-quarters in Port Clarence on the 28th of August, the *Dædalus* leaving for the south on the 1st of October.

The *Amphitrite*, Captain Frederick, with Comm^r. Maguire and a fresh crew for the *Plover*, arrived at Port Clarence on the 30th of June. Leaving on the 12th, Comm^r. Maguire proceeded to the north, and went up to Point Barrow in the boats, when a rapid survey of the harbour was made, and sufficient depth of water for the *Plover* was found, he returned to that vessel, and succeeded in placing her in winter-quarters at Point Barrow on the 21st of August, and was frozen in on the 24th of September. Before being frozen in he succeeded in his boats in searching the coast to the eastward as far as the Return Reef of Franklin.

The *Plover* did not get clear of her winter-quarters at Point Barrow until the 7th of August, and on the 11th of the same month fell in with the *Amphitrite*, Captain Frederick, and, after replenishing her provisions, returned to Point Barrow on September 7th. In attempting to prosecute the search easterly, an armed body of Indians of the Koyukun tribe were met with, and were so hostile that he was compelled to return; otherwise he would, in all probability, have reached the *Enterprise*, which vessel was stopped by the ice in Camden Bay, about 80 miles from the furthest point reached by him.

The *Plover* cleared her winter-quarters on July 19th, and, having received fresh supplies from H.M.S. *Trincomalee*, returned to Point Barrow on August 28th, being in no way impeded by the ice. The same evening the *Enterprise* arrived from the southward. This latter vessel, which had wintered in lat. 70° 8′, long. 145° 29′ w., 245 miles to the eastward of Point Barrow, where she was beset on September 16th, and frozen in on the 26th in the pack in Camden Bay, 4 miles from the shore.

On the 10th of July the whale-boat of the *Enterprise* under the command of Lieutenant Jago was despatched from her to Point Barrow, at which spot she arrived on the 24th. The ice broke up sufficiently to admit of the ship being moved on the 15th, and she reached Point Barrow on the 8th of August: Point Hope on the 10th; but, owing to the prevalence of southerly winds and a strong northerly current, did not arrive at Port Clarence until the evening of the 21st, when we communicated with the *Rattlesnake*, and found that the *Plover* had sailed for Point Barrow two days previously. After receiving some supplies, the *Enterprise* left for Point Barrow on the afternoon of the 22nd. On the 28th we made the ice in lat. 71°·0 and long. 159°·0 W., and reached Point Barrow the same afternoon, and, after communicating with the *Plover*, returned to Port Clarence on September 8th, the *Plover* arriving on the following day. Both vessels left for the south on the 16th.

In the course of the years 1865 and 1866 expeditions were equipped by Americans at San Francisco with the view of laying down a telegraph-cable across the continents of Asia and America. Mr. Whymper, who accompanied the expedition, has published an interesting account of his explorations, from which a few extracts have been made, as they bear on the ice movement.

In 1865 soundings were taken across Behring Sea between the 64° and 66° of latitude, when the bottom was found to be very even, with an average depth of 19½ fathoms.

In 1866 Mr. Whymper left Petropaulowski on August 6th, and reached Plover Bay on the 14th, where 14 men were left to pass the winter. Leaving Plover Bay on the 20th, Norton Sound was reached on the 24th. The ice in Norton Sound forms early in October, but is frequently broken up and carried to sea. On Christmas eve all the ice was blown out of the bay. In the spring the bay was not clear of ice until the third week in June.

He then proceeded overland to the Kwipak or Yukon River, and spent the winter at Nulato (the Russian fort where Lieut. Barnard was killed in 1851). Nulato by the river is 600 miles from its mouth; opposite the Fort it is 1¼ mile wide, and occasionally opens out into lagoons 4 and 5 miles across. The ice began to move on the river opposite the fort on April 5th, and on May 19th was rushing past at the rate of 5 or 6 miles per hour, bringing with it a large quantity of drift-wood, and rising 14 feet above its usual level. On the 26th Mr. Whymper left Nulato and, ascending the river 600 miles, reached Fort Yukon on June 23rd, which he estimates to be in latitude 66° N. This is the Hudson's Bay post to which the Rat Indians brought a communication from Commr. Maguire, with whom they fell in with near the mouth of the

Colville. They afterwards came on board the *Enterprise* just as she was on the point of leaving her winter-quarters in Camden Bay in 1854. The fort was established in the year 1847. Mr. Whymper left the fort on his return on July the 8th : the current sometimes carrying him 100 miles in 24 hours, they arrived at Nulato on the 15th, and on the 23rd entered the northern mouth of the river (the Aphoon), and reached Behring Sea the same afternoon.

The Kwipak, or Yukon, empties itself into Behring's Sea through five mouths, all of which are shoal, and extend from lat. 62° to 63° 25'. The Russian boats from Michaelowski usually occupy 35 days in ascending the river to Nulato; but that post can be reached in 5 days from the head of Norton Sound by travelling overland.

Extract from a letter from Captain Long to H. M. Witney, Esq., dated Honolulu, November 5th, 1867 :—

" Wrangell Land was first seen on the evening of the 14th of August, and the next day at 9.30 A.M. the ship was 18 miles distant from the west point . . . . which was found to be in lat. 70° 46' and long. 178° 30' E. . . . . The lower parts of the land were entirely free from snow, and had a green appearance as if covered with vegetation. . . . . Near the centre, or about long. 180°, there is a mountain which has the appearance of an extinct volcano : by approximate measurement I found it to be 2480 feet high. . . . . The south-east cape, which he named Cape Hawaii, was found to be in lat. 70° 40' N. and long. 178° 51' W. . . . . From long. 175° to long. 170° E. there were no indications of animal life in the water. It appeared almost as blue as it does in the middle of the Pacific Ocean, though there was but from 15 to 18 fathoms in any place within 40 miles of the land."

Captain Long thinks a propeller might readily have steamed up north on the west or east side of this land ; and he believes it to be inhabited. According to his track-chart he made Cape North on August 2nd, sailed along the Asiatic continent, passing close to Cape Iakan on the 4th, and reached Cape Chelagskoi on the 9th. On the 10th the furthermost western point was obtained in long. 170° 30' E. and lat. 70° 45' N.

Captain Rodgers, in 1855, reached the 72° of latitude in long. 174° 40' W., afterwards, returning southerly, he passed between Wrangell Land and Cape Jakan, having a depth of 25 fathoms water, and reached long. 176° 40' E. in lat. 70° 45' N.

Extract of a letter from Captain Craynor to Mr. Witney, dated November 1, 1867 :—

" On my last cruize I sailed along the south and east side of

Wrangell Island for a considerable distance three separate times,
and once cruized along the entire shore. . . . . I made the south-
west cape to be in N. lat. 75° 20' and E. long. 178° 15', and the
south-east cape in lat. 71° 10' and long 176° 40' w. . . . . The current runs to the north-west from 1 to 3 knots per hour. In long.
170° 10' w. we always find the ice-barrier from 50 to 80 miles
further south than we do between that and Herald Island. . . . .
In such shoal water the currents are changed easily by the wind."

Captain Long, in a letter dated January 15th, 1868, thus sum-
marises his opinion of the currents in Behring Straits:—

" The currents here have been found variable : in the spring and
summer the current is always setting towards the north ;
in the autumn and winter months, from information derived from
natives of the coast and whalers that have wintered in Plover
and St. Lawrence Bays, the current is found setting towards the
south. The barque *Gratitude* was wrecked in lat. 82½° N." (? 72½° N.),
" long. 168°, about 40 miles from Cape Lisburn, in the early part
of July, 1865, was seen in the month of August near Herald Island,
170 miles in a N.N.W. direction from the position in which she was
wrecked.

" The *Ontaria* was wrecked in September, 1866, in lat. 70° 25',
and during the following winter was seen by the natives drifting
through Behring Strait to the south, and was afterwards seen on
shore in lat. 64° 50' N.

The following account of the wreck and abandonment of the
whaling fleet off Wainwright Inlet, in September 1871, is taken
from the ' Hawaiian Gazette:'

" The fleet passed through Behring Straits between the 18th
and 30th of June. In July the main body of the ice was found
about lat. 69° 10', with a clear strip of water running to the north-
east along the land. In the second week in August most of the
ships were north of the Blossom Shoals, and some as far as Wainwright Inlet. Here they remained fishing until August 29th,
when a south-west wind set the ice inshore very fast, and at length
the ships were all jammed close together. On September 7th the
barque *Roman* was crushed by the ice like an eggshell, in forty-five
minutes, and on the 8th the barque *Awashonks* was crushed. On
the 9th the weather was calm, and the water around the ships froze
over. Not having provisions to last over three or four months, a
meeting of the Masters was held on the 13th, when it was determined to abandon the ships, which was done at 4 P.M. on the 14th,
and reached the barques *Arctic*, *Midas*, and *Progress*, on the 16th ;
the distance traversed in the boats being about 70 miles. In all,

thirty-one vessels were either crushed or abandoned, and seven vessels were saved.

Dr. Simpson's Remarks.—" Through the large opening between the American and Asiatic continents, occupied by the Aleutian Islands, there is an almost imperceptible set from the Pacific Ocean northwards, the waters of which, retaining the impulse given them by the earth's rotation in a lower latitude, draw towards the American shores, and throw themselves into Norton Bay. They are thence driven with increasing force along the coast of America opposite the island of St. Lawrence, diffusing themselves to the north of that island to be carried with lessened speed through the Straits of Behring, after receiving in the latter part of their course the freshwater stream falling through Grantly Harbour into Port Clarence.[1] Spreading again over a larger space, they receive a further tribute from Kotzebue Sound, which is very palpable off Port Hope. Again in the latitude of Icy Cape the earth's rotation gives them an easterly set, forming an almost constant current along the north coast of America to Point Barrow, whence it pursues a direction north-east. Throughout all this course the current is subject to retardations, and even surface-drifts in an opposite direction, caused by northerly and north-easterly winds, but it is also accelerated by southerly and south-westerly gales."

" In the beginning of the summer the eastern side south of the straits is free from ice, and Norton Bay itself is usually cleared as early as April. After the middle of June not a particle of ice is to be seen between Port Spencer and King's Island; whilst the comparatively still water north of St. Lawrence Island is hampered with large floes until late in July."

" This can be satisfactorily accounted for by the existence of a northerly current partly driving and partly throwing the ice down from the American shores. There is scarcely a particle of driftwood to be had on the Asiatic coast from Kamschatka to East Cape, whilst abundance is to be found in Port Clarence and Kotzebue Sound, as well as along the whole American shore from Norton Bay to Port Barrow.

" Although it has been found that pine-trees 60 inches in girth grow here on the banks of the American rivers, within the 67th parallel of latitude, yet from the frequently larger size of the trunks and their great abundance, it is evident these northern regions, including Norton Bay, cannot supply the quantity: and more southern rivers, whether Asiatic or American, or both, must

[1] Dr. Simpson was not aware of the importance of the River Yukon.

be looked to for the numerous multitude of water-worn stems and roots strewed almost everywhere along the beach. Their southern origin would also seem to be indicated by the presence in many of them of the remains of the *Teredo navalis*, which could hardly retain life throughout the rigour of eight or nine months' frost every year." It would seem that between St. Lawrence Island and the coast of Asia the current is variable, and seldom entirely free from ice until late in July; hence the many disasters to whalers in 1851, and the difficulties the *Dædalus* and *Enterprise* encountered the same season by taking the westward passage, whilst an open boat from the *Plover* was able, between the 17th of June and the 1st of July, to make the run to Michaelowski in Norton Bay and back without her crew seeing any ice."

"The *Amphitrite* in 1852 was able to reach Port Clarence on the 30th June by the eastern passage without seeing but one floe, which had probably been recently released from some of the nooks in Norton Bay: although late in the same month the master of a whaling ship reported that the ice was still fast as low as lat. 58° and 60° between the longitude of Gore's Island and the coast of Kamschatka."

"To the northward of Cape Prince of Wales the warm water is always found on the American coast. From frequent observations the temperature of the water near East Cape was found to be 35°, while that near Cape Prince of Wales was 53°. The cold current sets south along the coast of Asia."

"From recorded observations it appears that the coast from Icy Cape to Point Barrow is frequently packed with ice in the end of July and the beginning of August. The cause of this seems to be the occasional prevalence of westerly and north-westerly winds, which drive the pack upon the coast, again to be cleared away by the north-east current along shore as soon as these winds have spent their force: and southerly and south-east winds will have the opposite effect of driving it in a more northerly direction, and leave the navigation more open than usual. At Icy Cape the current on Captain Beechy's chart is marked running both ways along shore, but not, it is presumed, with the regularity of a regular tidal ebb and flow. During the continuance of an easterly gale from the 29th of July to the 5th of August, and a fresh breeze following for two days at that cape, floating substances were observed to drift slowly to leeward, whilst the waves were short, irregular, and much more broken than usual, to a distance of 12 miles off, as if caused by a weather-current. This may, however, be partly owing to the shoals extending 4 miles off the land. On the 3rd, a whaling vessel stood within 6 miles of the shore, tacked, and stood out again, making such progress to windward as a sailing vessel could only do when favoured by a strong weather-current."

" From Icy Cape to the Seahorse Islands, in addition to drift-wood, there is strewed along the beach a quantity of coal, which, though much water-worn, may, in some of the indentations, be collected in sufficient abundance, and bituminous enough to make an excellent fire for cooking. It is of the sort called candle-coal, and some of the pieces are sound enough to be carved by the natives into lip ornaments."

" At the Seahorse Islands it is found as fine as small gravel, and, on digging into the beach, is seen to form alternate layers with the sand; but between Wainwright Inlet and Icy Cape it is gathered in knots of a convenient size for fuel. This may be taken as a farther evidence of the set of the current, as the nearest known point whence the coal is brought is that marked on the chart as Cape Beaufort. The whole extent of the coast from below Icy Cape to Point Barrow is bordered by a beach of gravel, which has likewise a southern origin, and determines the form of the continent, offering as it does an effective barrier to the encroachment of the sea, which would otherwise speedily undermine the earth-cliffs behind. All that can be seen from the seaboard landward is a flat, alluvial plane, seldom exceeding 20 feet in elevation, and containing numerous pools and lagoons of fresh water, but without a tree or bush to relieve the view."

" The tides are hardly appreciable and very irregular at Kotzebue Sound and Port Clarence; there the sea usually retains a very low level during the prevalence of northerly, north-easterly, and easterly winds, and the highest levels occur with southerly and south-westerly gales. During a stay of seven days at Icy Cape, with a prevailing gale at east and E.N.E., the same low-water level obtained as much as 4½ feet below the highest surf-mark, the undeniable effects of westerly and south-westerly winds. With the drifted material left on those marks where the shore has a westerly aspect were several varieties of dead shells, identical in species with those previously dredged from the bottom of the sea in deep water, 25 to 30 fathoms in the straits and north of them."

It will be seen by the foregoing abstracts that the navigation of the Arctic Sea between Behring Straits and Point Barrow is comparatively easy to vessels fitted for ice-navigation. The current of warm water from the Pacific sets continually to the north-east throughout the summer months, and forms a lane between the pack and the land which enabled the *Blossom's* barge, on August 21st,

1826, to reach Point Barrow. Mr. Shedden, in a schooner-yacht of 140 tons, rounded the Point on August 4th, 1849.

The *Investigator*, Comm^{r.} McClure, on August 5th, 1850.
The *Enterprise*, Captain Collinson, on August 20th, 1850.
    ,,   ,,   ,, on July 25th, 1851.
The *Plover*, Comm^{r.} Maguire, on August 20th, 1852,
and wintered there, being frozen in on September 24th.
The *Plover* left her winter quarters on August 7th, 1853.
Returned to   ,, · ,, on September 7th, 1853.
Left again her   ,,   ,, on July 19th, 1854.
Returning to her   ,,   ,, on August 28th, 1854.
Enterprise returning from the eastward,
Rounded the Point   on August 8th, 1854.
And returned from Port Clarence on August 28th, 1854.

The season of 1854 was, undoubtedly, the most open, the ice being so far from the Point that the whaling ships were enabled to fish off it.

The season may be considered to be open from the beginning of July to the middle of September. The pack is usually met with off Icy Cape, and should westerly winds have prevailed and forced the pack into the shore, a vessel will do well to wait until the wind subsides, when the current will be sure to open the lane between the land and the pack. Easterly winds check the current, and, after a continuation of them, there is a set alongshore to the southward. Some natives got adrift in the ice in 1853, and were carried by this set to the southward of Icy Cape, the land being always in sight.

In both years the *Plover* wintered at Point Barrow. The ice round the Point was broken up, and swept to the northward by south-westerly gales. At times no ice could be seen from the mast-head. In 1853 this disruption occurred in December, and caused the water to rise 3½ feet above the highest spring-tide. The temperature at the same time rose to + 30° F. In January, 1854, the same thing occurred, the thermometer on this occasion rising to + 27°. During both winters a water-sky to the north-west was generally observed from the ship, unless after a long continuance of north-westerly winds or calm weather. There is but little rise and fall of the tide, 0·7 inches being the average. With fine weather or easterly winds they were very regular, but a south-west gale upset them altogether.

Eskimo whale-fishing commenced on May 7th, 1853, the open

water being 4 miles from Point Barrow, extending in an E.N.E. and w.s.w. direction, with a depth of 10 fathoms water.

Between the 4th and 7th of July about thirty oomiaks, carrying about 150 people, went to the eastward. The ship swung to her anchor on July 25th, and the ice was in motion in the offing on July 30th.

An abstract from the *Plover's* log, which I have to thank Staff-Comm' Hull for, shows the number of days in each month that open water, as well as a water sky, was seen from that vessel during the two winters spent at Point Barrow. In 1852-53, open water was seen on twenty-seven days between October and April, and in 1853-54 on seven days only, whilst the indication of open water occurred during the same period in 1852-53 on fifty-seven days, and in 1853-54 on sixty-two days. December, January, and February appear to be the months during which the ice is more frequently in motion.

It will be, perhaps, advisable here to introduce the tables of monthly temperatures taken from Dr. Simpson's paper.

Month.	1852-53.			1853-54.		
	Max.	Min.	Mean.	Max.	Min.	Mean.
Sept. .	+42·	+12·	+28·9	+41·	− 3·	+23·0
Oct. .	+27·	−21·	+ 3·9	+14·	−22.	− 0·8
Nov. .	+25·	−37·	− 7·8	+22·	−26·	− 7·4
Dec. .	+28·	−37·	− 8·7	+ 7·	−40·	−18·7
Jan. .	+16·	−43·	−23·8	+27·	−37·	−13·7
Feb. .	+ 3·	−36·	−17·4	− 3·	−45·	−27·9
March	+24·	−37·	−12·7	+23·	−42·	−17·8
April.	+33·	−40·	+ 3·8	+26·	−17·	+ 1·6
May .	+44·	− 6·	+18·5	+42·	− 3·	+20·5·
June .	+45·	+17·	+32·1	+47·	+24·	+32·8
July .	+52·	+26·	+35·4	+51·	+28·	+37·2
August	+49·	+31·	+38·7	+48·	+29·	+39·1
Means	+32·3	−14·2	+ 7·2	+28·7	−12·8	+ 5·7

It is remarkable that, though the winter of 1852-53 was warmer than the ensuing one, the *Plover* was detained by the ice in her winter-quarters until August 7th; whereas in 1854 she made her escape to the southward on July 23rd, and the ice during the summer was so far off the Point, that the whale-ships fished off it. The temperatures observed on board the *Enterprise*, which vessel wintered in the pack 245 miles to the eastward, in 1853-54, are given for the sake of comparison.

Month.	Max.	Min.	Mean.	Thickness of ice on the 1st.
				inches.
October	+24	−20	+ 0·6	0·07
November	+20	−33	− 9·6	2·02
December	− 4	−51	−26·0	2·11
January	+27	−49	−16·2	4·00
February	− 5	−51	−31·8	5·00
March	+16	−47	−20·0	6·00
April	+19	−26	−00·9	6·02
May	+47	..	+23·0	7·00½
June	+46	+26	+32·4	7·02[1]
July	+53	+27	+37·5	4·11

On July 10th the water along the coast was sufficiently open to send the whale-boat to Point Barrow. On the 15th the ice broke up, which was three days earlier than at Point Barrow. The ship left Camden Bay on the 20th, but, owing to obstruction by the ice, did not reach Point Barrow until the afternoon of August 7th.

In comparing the monthly temperatures of Camden Bay and Point Barrow, the increased temperature at the latter place is very perceptible, and is, no doubt, occasioned by the open water. Neither open water nor water-sky was seen from Camden Bay. In the month of April an attempt was made to go north from the *Enterprise* with three sleighs; but on the second day the hummocks were found to be impassable. One sleigh utterly broke down, and several accidents from severe falls rendered it necessary to give up the attempt and return to the ship. The snow-drift on these hummocks lay in continuous ridges east and west, indicating that no dislocation of the ice had taken place during the winter.

The condition of the ice north of the American continent affords a remarkable contrast with that on the Asiatic shore, where year after year open water is found all the way from Kotelnoi Island to North Cape, a distance nearly 1000 miles. Here, instead of a compact pack, which in the neighbourhood of Point Barrow is occasionally moved off the shore and brought back by the force of the wind, but which appears to remain perfectly quiescent along the coast to the eastward, the water, for some reason or other, is prevented from freezing, and the traveller is continually brought to a stop by the thinness of the ice or open water itself. This water, on referring to Baron Wrangell's and M. Von Anjou's Journals, will be found to be always in motion, and remarks such as follows

---

[1] On the 15th, 6·3 inches.

are found in their Journals :—" Current ½ a knot in an E.S.E. direction." " Strong current running E.S.E." " Off Schalarov Island, current running 1½ knot to the eastward." " Though the wind was westerly the pieces of ice drifted from east to west ; the sledge-drivers were of opinion that this was the ebb tide, the regular six-hourly return of which they had noted."

It will be seen on reference to the meteorological register kept on board the *Enterprise*, where the thickness of the ice was measured on the first of every month, that the thickness increased up to June 1st, the mean temperature of the month of May being + 23°. The change in the character of the ice cannot therefore be ascribed to the temperature of the atmosphere, but will probably be found due to the motion of the water. These Polynias, or open spaces of water, have since been fallen in with to the north of Grinnell Land and in the upper portion of Smith Sound, and they were seen by Lieutenant Payer in the recent voyage of the *Tegethoff*, as far north as the 82° of latitude. It is to be noted that the open water on the Siberian coast occurs in comparatively shallow water under 20 fathoms, whereas in the neighbourhood of Point Barrow the water deepens with great rapidity.

NATIVE NAMES FOR SOME PLACES BETWEEN THE MACKENZIE RIVER AND POINT HOPE.

The Mackenzie,	*Iuna* (?)
Village between it and Point Kay,	*Pe-ock-rhu.*
Point Kay,	*Te-kre-ra.*
Reef East of Herschel Island,	*Ke-yuk-ta-zia.*
Herschel Island,	*Ke-yuk-ta-hue.*
Barter Island,	*Noo-na-niaou.*
Fishing-station this side,	*Ac-hut.*
Village visited by us in the autumn,	*Noo-na-ma-luk.*
Village about 7 miles S.E. by E. from ship,	*Noo-iroo-a.*
Romanzoff chain of hills,	*Chud-loo-o-sak.*
Canning River,	*Kook-Doak.*
Flaxman Island,	*Kapa-gill-luk.*
Between Point Barrow and Flax- man Island,	*Che-gea.*
Point Barrow,	*Noo-icook.*
Beyond Point Barrow,	*Ot-kia-mik-miot.*
Northern stream between Refuge Inlet and Cape Smyth,	*Oo-fi-la.*
Cape Smyth,	*Noo-oo.*
Refuge Inlet,	*Noo-naboo* or *Il-lip-su.*
Inlet south of it,	*Too-na-mut.*
Cape Lisburn,	*Te-ga.*

K

Village between Cape Dyer and Cape Lisburn,	Woo-ut.
Village north of Asses' Ears,	Ka-na-due.
Point Hope,	Noo-na.
Icy Cape,	Ol-ron-na.
Wainwright Inlet,	Kog-ru-ak.

————————

EXTRACTS FROM THE LOG OF H.M.S. *Plover*, 1852.

On passage from Port Clarence to Point Barrow, encountered the ice off Icy Cape, but found no difficulty in reaching Point Barrow, where we anchored on the 3rd September.

September.
    4. A heavy N.W. gale brought in the pack.
    9. An easterly wind cleared off the same.
    13. Pack returns with a N.W. wind.
    15. Pack cleared out by a southerly wind.
    20. N.W. winds bring the pack in.
    22. Open sea beyond the ground hummocks at 5 miles N.W., *true*, of Point Barrow.
    26. Frozen in.

————————

October.
    21. Open water 5 miles from Point Barrow.
    26.
    27. } Water Sky.—S.W. to N.N.E.
    28.

————————

November.
    4.
    5. } Water Sky.—S.W. to N.N.E.
    7. Open water within ½ mile of Point Barrow.
    8. Water Sky.—W.S.W. to N.E.
    10. Water Sky.—S.W. to N.
    22.
    24.
    27. } Water Sky to the N.W.
    30.

       Water Sky seen for 9 days in November.

————————

December.
    11.
    12. } Water Sky.—W.S.W. to E.
    13.
    15. } Water Sky.—S.W. to N.W. and N.
    16.
    17. Break up of the ice with a heavy S.W. gale. Thermometer + 30° Fahr. From this date until the 1st of January the sea may be said to have been open for another southerly gale. On the 28th took all the young ice out to sea.

       5 days Water Sky, and open water on 16 days in December.

1853.

January.
  1. A few hummocks in sight.
  2. Sea freezing again.
  5. Water Sky.—W. to N.E.
  11.⎫
  12.⎪
  13.⎪
  14.⎪
  15. ⎬ Water Sky.—W. to N.E.
  16.⎪
  17.⎪
  18.⎪
  19.⎭
  24.⎫
  25.⎪
  28.⎬ Water Sky.—S.W. to N.E.
  29.⎭

          Water Sky seen on 16 days in January.

February.
  4. Water Sky.—N.W. to E.
  7.⎫
  8.⎭ Open water seen from Point Barrow.
  9. Water Sky.—W.N.W. to N.N.E.
  12. Ice packed heavily on W. side of Point Barrow, piled to the height of
20 feet.
  15.⎫
  17.⎭ Water Sky.—W.S.W. to N.N.E.
  19.⎫
  20.⎬ Open water again seen from Point Barrow, after a strong easterly gale.
  21.⎭
  23.⎫
  25.⎪
  26.⎬ Water Sky.—W. to N.
  27.⎭
      Water Sky seen on 9 days, and open water on 5 days in February.

March.
  1. Water Sky.—W. to N.
  7. ⎫
  8. ⎪
  9. ⎪
  13. ⎬ Water Sky.—W. to N.E.
  14. ⎪
  25. ⎪
  26. ⎭
          Water Sky seen on 8 days in March.

April.
  3. Water Sky.—W.S.W. to N.E.
  7. Water Sky.—N. to N.E.

                                                          K 2

April
18.
23.
24.
25.
26. } Water Sky.—W. to N.E.
27.
28.
29.
30.

Water Sky seen on 11 days in April.

May.
4.
5. } Water Sky.—N.W. to N.E.
6.
7. } Open water seen from Point Barrow after strong easterly winds.
8.
to } Water Sky continuous.—N.W. to E.
31.

Water Sky seen on 25 days in May.

June.
1.
2. } Water Sky.—W.N.W. to N.E.
3.
12.
to } Water Sky.—W. to N.E.
19.

Water Sky seen on 11 days in June.

July.
9. Boats left for the open water.
10. Open water seen from ship.
24. Ship free from ice, but pack close in to the Point.
30. The grounded hummocks off Point Barrow moved to the N.E.
31. Open water off Cape Smyth, south of Point Barrow.

August.
7. Pack left the land: *Plover* left Point Barrow.
8. Beset in Peard Bay ; current running N.E.
9. Cleared the pack off Cape Franklin.

August.
31. On return to Point Barrow met the pack 15 miles north off Icy Cape ; current setting N.E.

September.

2.
to } Beset off Refuge Inlet.
5.

6. Cleared the pack, but again beset off Point Barrow, and carried to the N.E. at the rate of 2 miles an hour. Succeeded in getting alongside of a grounded hummock.

7. Cleared pack, and anchored in Point Barrow.

16. Frozen in.

25. Inshore waters frozen; but open sea from Point Barrow.

October.

3.
4.
6.
7.
11. } Water Sky.—W. to N.E.
12.
13.
14.

Water Sky seen on 8 days in October.

November.

3.
22. } Water Sky.—N.W. to N.E.
23.

Water Sky seen on 3 days in November.

December.

7. Water Sky.—W.N.W. to N.E.

8. Great pressure of ice on outer spit; ice forced up 22 feet. No apparent cause. Wind S.W. and calm.

9.
10.
11. } Water Sky.—N.W. to N.E.
12.
13.

21.
22.
23. } Water Sky.—N.W. to E.N.E.
27.

Water Sky seen on 11 days in December.

1854.

January.

2.
3. } Water Sky.—N.W. and N.
5.

12. Heavy S.W. gale taken out the ice from the Point.

13. Open water from mast-head as far as could be seen.

14. Ditto do. Temperature + 28° Fahr.

15. Sea freezing.

January.

16. Open water off Point Barrow.

17.
18. } Water Sky.—N.W. to E.

19.
20. } Ice piled to the height of 30 feet on the spits S. of Point Barrow.

Water Sky seen on 12 days in January, and open water on 4 days.

February.

23.
24.
25. } Water Sky.—N.W. to E.

26.
27. } Open water seen from the grounded hummocks near Point Barrow.

Water Sky seen on 5 days in February.

March.

2.
4. } Water Sky.—N. to E.

5.
6.
7. } Water Sky.—W. to E.

12. Captain Maguire walked to the edge of the shore floe, about 10 miles to the N.W. of Point Barrow; found the ice in the open water to be setting slowly to the eastward.

20.
21.
23.
25. } Water Sky from N. to E.
26.
27.
28.

Water Sky seen on 13 days in March.

April.

6.
9.
10.
12.
13.
14.
15.
16.
17. } Water Sky.—N.W. to N.E.
19.
20.
22.
23.
26.
27.
29.
30.

Water Sky seen on 17 days in April.

Water Sky was seen all May. Officers away with natives whaling.

May.
13. Water only 2¼ miles from Point Barrow. Loose ice on that day moving slowly to the southward. Water Sky seen all the month.

June.
All June a Water Sky observed.

July.
10. Open water at Point Barrow.
15. Ship free from ice.
18. General break up.
23. *Plover* cleared the pack-ice off Wainwright Inlet.

August.
19. *Plover* sailed from Port Clarence to Point Barrow without being in any way impeded by the ice.
30. Sailed South from Point Barrow.

The above notes, which show the prevalence of open water in the vicinity of Point Barrow, where H.M.S. *Plover*, Commander Rochfort Maguire, wintered in 1852-3-4, are copied from that vessel's log-book.

4th March, 1875.   THOMAS A. HULL.

2.—A SHORT ACCOUNT OF THE EXPLORATION OF THE POLAR SEA

Between POINT BARROW and the RIVER MACKENZIE, including the VOYAGES of the *Investigator* and *Enterprise* to BANKS LAND.

Voyage of Mackenzie to the Polar Sea, 1789.—Sir A. Mackenzie, attended by a German, four Canadians, and three Indians, together with two Canadian and two Indian women, left Fort Chipewyan on June 3rd, 1789, in four birch-bark canoes. The Slave Lake was reached on the 9th, where they had to remain six days to enable the ice to give way. They then entered at the west end of the lake the river which now bears the name of Mackenzie, and eventually reached the Great Northern Ocean on the 15th of July. Returning by the same route, the party regained Fort Chipewyan on September 12th.

Captain Franklin's Second Voyage, 1825-26.—Three boats were built at Woolwich for this expedition, one of which was 26 feet, and the two others 24 feet long, and a small vessel, 9 feet long, 4 feet 4 inches wide, which weighed only 85 lbs., and could be made up in five or six parcels. These were forwarded to York Factory in 1824.

The expedition, consisting of Captain Franklin, Lieutenant Back, Dr. Richardson, Mr. Kendall, and Mr. Drummond, with four marines, left Liverpool in February, 1825. Passing through the United States and Upper Canada, Fort William, on Lake Superior, was reached on May 10th, and the Methye River on June 29th, where they joined the boats which had been forwarded from Hudson Bay, and arrived at Fort Chipewyan on the 15th of July. Leaving it on the 25th, Fort Resolution was reached on the 29th, and the Mackenzie River on the 3rd of August. Quitting Fort Simpson on the 5th, they arrived at Fort Norman on the 8th, and Fort Good Hope on the 10th, and the Polar Sea on the 16th, and returned to Fort Good Hope on the 23rd; arrived at Great Bear Lake on September 1st: the total distance travelled over from New York being 5803 miles.

Passing the winter at Fort Franklin, in lat. 65° 12', long. 123° 13', Captain Franklin, accompanied by Lieutenant Back, in the two boats which were named the *Lion* and *Reliance*, left the Fort on June 22nd, 1826, arrived at Fort Norman on the 25th, Fort Good Hope on July 1st. The mouth of the river was reached on the 7th. The Eskimo were met with, who attempted to pillage the boats.

Detained by the ice, being pressed close on the shore, but little progress was made. The rise and fall of the tide was found to be about 2 feet. Point Kay was reached on the 15th, Herschel Island on the 17th, and Point Demarcation on the 31st. A black whale and several seals were seen, and the ice was driving with great rapidity to the westward. Barter Island was arrived at on the 4th of August, and here a musket was left by accident on the beach; this musket was seen at Point Berens in 1850, by Lieutenant Pullen. On the 6th they got to Flaxman Island; on the 7th and 8th, at Lion Reef, the tide was found to be regular, rising 16 inches. After great obstruction, Point Anxiety was passed on the 16th, when further progress to the west was found to be impracticable this season. Returning to the east, Flaxman Island was gained on the same day that Mr. Elson, in the barge of the *Blossom*, reached Point Barrow from Behring Straits, August 22nd, being 160 miles distant from Captain Franklin's furthest point. Demarcation Point on the 24th, Herschel Island on the 26th, and Garry Island, at the mouth of the Mackenzie, on the 29th; and by aid of the tracking-line, Fort Good Hope on September 7th, and Fort Franklin on the 21st.

The distances traversed are as follows:—

	Miles.
From Fort Franklin to Point Separation	525
„ Point Separation to Pillage Point	129
„ Pillage Point to Return Reef	374
„ Return Reef to Fort Franklin	1020
Total	2048

Boat Voyage of Messrs. Dease and Simpson from the River Mackenzie to Point Barrow in 1837.—Leaving Fort Chipewyan in two clinker-built boats of 6 feet beam and 24 feet keel on June 1st, Messrs. Dease and Simpson were detained prisoners by the ice at Fort Resolution from the 10th to the 21st; they passed the Hay River on the 23rd, and arrived at Fort Simpson on the 28th, and at Fort Norman at 10 P.M. on July 1st, having travelled 250 miles in 48 hours; and reached Fort Good Hope on the evening of the 4th. Starting again on the 5th, Eskimo caches were reached on the 8th, and on the following day the natives themselves were met with, and the Arctic Ocean reached. Detained by a north-west gale at Shingle Point, Point Kay was passed on the afternoon of the 11th. The violence of the wind prevented their moving until the 14th, when the first regular flow and ebb was observed, and taking advantage of the opening in the ice, they passed inside Herschel Island. Some bones of an enormous whale were found here. On the 15th Demarcation Point was reached. The tide, though insignificant, did us good service. Flaxman Island was gained on the morning of the 20th; detained by a gale on the 21st and 22nd, they reached Return Reef on the evening of the 23rd. Strong gales delayed them at Point Comfort until the 26th, when Harrison Bay was crossed. At Cape Simpson the tide rose 10 inches. On August 1st Mr. Simpson started on foot with five men, each carrying from 40 to 50 lbs. After passing Port Tangent 10 miles, they obtained an oomiak from the Eskimo, in which they crossed Dease Inlet and got to Point Christie on the 3rd, and gained Point Barrow on the 4th; thus connecting the discoveries of Beechey with those of Franklin, and perfecting the outline of the American continent from the 156th to the 108th meridian. Returning easterly, the boats were reached on the 6th. Mr. Dease had ascertained the rise and fall of the tide to be 15 inches, and that the flood came from the north-west. Demarcation Point was reached on the 11th, where an easterly wind detained them until the 15th; and it was not until the evening of the 17th that Tent Island was reached. The ascent of the Mackenzie was performed almost exclusively by towing, at the rate of from 30 to 40 miles per day, and Fort Good Hope arrived at on the 28th.

Lieutenant Pullen's Boat Voyage from Point Barrow to the Mackenzie River.—Lieutenant Pullen left the *Plover* off Wainwright Inlet on the 25th of July, 1850; and in company with two other boats, and Mr. Shedden's yacht, the *Nancy Dawson*, reached Point Barrow on August 2nd. Here the escort left them, and the two boats proceeded along the coast.

Point Pitt was passed on the 7th.
Cape Halkett   ,,   9th.
Point Berens   ,,   11th.
Lion Reef on the 14th. Many seals; rise and fall of the tide 18 inches; current strong to the west; wind fresh, north-east.
Flaxman Island on the 16th.
Manning Point   ,,   18th.
Humphrey Point   ,,   20th. Two whales seen.
Herschel Island   ,,   22nd. Yellow water.
Entered the Mackenzie River on the 27th. Tracks of bears, moose, and reindeer frequent.

At Fort Macpherson, September 5th. Arrived at Fort Norman, October 6th. The ice in the Mackenzie set fast on November 12th. Snow-birds arrived on the 24th of April, 1851 ; ducks, May 4th. Ice began to break up on May 14th. Left Fort Simpson, July 11th ; at Fort Good Hope, 16th ; Port Separation on the 20th ; and the Arctic Sea on the 22nd ; Richard Island, 24th ; Cape Dalhousie, August 3rd ; and Cape Bathurst, August 10th. Found the ice packed close on the shore ; small whales seen. The boats remained here until the 15th, and on the 30th Captain M'Clure landed here from the *Investigator.*

Garry Island was reached on the 26th.
Fort Macpherson   ,,   7th of September.
Fort Good Hope   ,,   17th   ,,
Fort Simpson   ,,   5th of October.

Voyage of H.M.S. 'Investigator' from Point Barrow to the Bay of Mercy.—The *Investigator* rounded Point Barrow at midnight on August 5th. Reached Point Drew on the 8th ; Jones Island on the 11th. Ran on a shoal 8 miles north of Yarborough Inlet on the 14th. On the 15th the ice closed in from the north ; anchored to await some favourable change. The ice eased off on the following morning, and the ship was warped through a lane 150 yards wide. The Pelly Islands were reached on the 21st. The temperature of the sea on reaching the coloured water of the Mackenzie rose from 28° to 39°. On the 30th reached Cape Bathurst, and communicated with the natives. On September 6th discovered Baring Land. Landed and took possession on the 7th. On the 11th the ship was beset in lat. 72° 52', and long. 117° 3'; but the ice continued in motion until October 8th, and the ship narrowly escaped destruction several times; on one occasion listing the ship 34°, when they were firmly fixed for the space of nine months in lat. 72° 47' N., and long. 117° 34' W., 4 miles from the Princess Royal Isles. Here three months' provision and a boat were deposited. On the 21st Captain M'Clure started with sleighs, and reached the entrance into Barrow Strait, in lat. 73° 30', and long. 114° 14' W., and thus established the existence of a north-west passage. On July 14th, 1851, the ice opened without any pressure; but the ship was so surrounded by it that they were only able to use their sails twice until August 14th, when they attained the furthest northern position in Prince of Wales Strait, viz., lat. 73° 14' N., long. 115° 32' 30″.W. Finding the passage into Barrow Strait obstructed by north-east winds setting large masses of ice to the southward, which had drifted the ship 15 miles in that direction during the last 12 hours, bore up and passed to the southward of Baring Island.

August 20th, lat. 74° 27' N., long. 122° 32' 15″ W. Have had clear water to reach thus far, running within a mile of the coast the whole distance, when progress was impeded by the ice resting upon the shore: secured the ship to a large grounded floe-piece in 12 fathoms.

August 29th, ship in great danger of being crushed or driven ashore by the ice coming in with heavy pressure from the Polar Sea, driving her along within 100 yards of the land for half a mile, heeling her 15°, and raising her bodily 1 foot 8 inches, when we again became stationary and the ice quiet.

September 10th. Ice again in motion, and ship driven from the land into the main pack, with a heavy gale from south-west. On the following day they succeeded in getting clear of the pack, and secured the ship to a grounded floe in 74° 29' N., long. 122° 20' W.

September 12th. Clear water along shore to the eastward. Worked the ship in that direction, with several obstructions and narrow escapes from the stupendous Polar ice, until the evening of the 23rd, when they ran upon a mudbank, having 6 feet water under the bow and 5 fathoms astern; hove off without any damage. Finding a well-sheltered spot upon the south side of this shoal, ran in, and anchored in 4 fathoms, in lat. 74° 6', long. 117° 51', on the 24th, and were frozen-in the same evening.

On October 4th, Mr. Court was sent to connect the position of the ship with the Point reached by Lieutenant Cresswell in May, which was distant only 18 miles. He reported open water a few miles from the shore.

On April 11th, 1852, Sir R. M'Clure proceeded to Melville Island, and reached Winter Harbour on the 28th, and returned to the ship on the 9th of May.

On August 10th, lanes of water were observed to seaward, and along the cliffs of Banks Land there was a clear space of 6 miles in width, extending along them as far as the eye could reach. On the 12th the wind, which had been for some time to the north, veered to the south, which had the effect of separating the sea-ice from that of the bay entirely across the entrance, but shortly shifting to the north, it closed again, and never after moved. On the 20th the temperature fell to 27°, when the entire bay was completely frozen over. During this summer the sun was scarcely seen, and Captain M'Clure states in his Journal: "nor do I imagine that the Polar Sea has broken up this season." On the 24th of September, the anniversary of their arrival in Mercy Bay, the thermometer stood at 2°, with no water in sight, whereas they entered the bay with the thermometer at 33°, and not a particle of ice in it.

On April 7th, 1853, Lieutenant Pim reached the *Investigator* from the *Resolute*; Captain M'Clure left that vessel on the same day, and reached the *Resolute* on the 19th.

Lieutenant Cresswell left the *Investigator* on April 15th, and reached the *Resolute* on May 2nd, and the *North Star* at Beechey Island on June 2nd.

Sir R. M'Clure's Remarks.—The currents along the coasts of the Polar Sea appear to be influenced in their direction more or less by the winds, but certainly on the west side of Baring Island there is a permanent set to the eastward, at one time we found it as much as two knots during a perfect calm; and that the flood-tide sets from the westward we have ascertained beyond a doubt, as the opportunities afforded during our detention along the western shore of this island gave ample proof.

The prevailing winds along the American shore and in the Prince of Wales Strait we found to be north-east, but upon this coast from south-south-west to north-west. A ship stands no chance of getting to the westward by entering the Polar Sea, the water alongshore being very narrow and wind contrary, and the pack impenetrable, but through Prince of Wales Strait, and by keeping along the American coast, I conceive it practicable.

Voyage of the 'Enterprise.'—After rounding Point Barrow in the pack, the *Enterprise* got into the land-water on July 31st, 1851, the edge of the pack was found to be in 7½ fathoms of water ; the temperature of the sea rose immediately from 32° to 37°, and reached as high as 46° during the day. Working to the eastward between the pack and the shore, which was sometimes as little as 3 and occasionally as much as 8 or 9 miles wide, as the River Colville was approached the colour of the water changed, and the main body of the ice was as far as 10 and 12 miles from the land. After passing the mouth of the Colville the land-water became strewed with large floe-pieces, rendering it difficult to beat to windward, and at length on August 5th we were compelled to make fast to a floe.

On reaching Lion Reef drift-wood was seen on the beach in great abundance, the current was here found to run w. by n. (true) 0·5 per hour.

Barter Island was passed on the 7th. The main body of the ice was found to be pretty close to Point Manning.

On August 8th the current ran to the N.N.E. 0·5 per hour ; great difficulty was experienced in steering the ship even with the boats ahead. The ice was much farther from the shore, and on the afternoon of the 9th we were in 17 fathoms water, and passed through a stream of drift-wood trending N.N.W. and S.S.E. The current at the surface ran E. by S. (true) 0·5, and at 10 fathoms N.N.E. 0·2 per hour. The temperature of the sea rose to 49°. On the 10th, at a distance of 28 miles from the land, a depth of 28 fathoms was obtained, and the current was found to set W.S.W. 0·7 per hour.

On the 13th of August Herschel Island was seen. Standing off shore on the 16th no bottom was obtained with 140 fathoms of line. On the 18th several streams of drift-wood were passed through, and one tree, 68 feet long, picked up. The edge of the ice trended N.N.E. and S.S.W. The current was found as follows :—

At 2.30 A.M., E. by N. (true) 1·0 knot per hour.
At noon,   N. by E.   ,,   0·5   ,,   ,,
At 5 P.M ,   S. by W.   ,,   0·7   ,,   ..

On the 20th the Pelly Isles were seen, and two islands to the E.N.E., in lat. 69° 37', and long. 134° 32', and in lat. 69° 39', and long. 134° 10'.

At 6.30 p.m.   the current set w. by s. ½ s. 0·6 knot per hour.
At 11.0 p.m.   „   „   w. by s.   0·4   „   „
At 2.0 a.m. 21st „   „   w. by n.   0·3   „   „

On the 24th we stood in towards Cape Brown, getting 5 fathoms
water 2 miles from the beach; on reaching off 34 miles we could
trace the pack from e.n e. round by north to s.w.

On August 25th[1] land was seen to the north, and at noon on the
27th, in lat. 71° 27', and long. 120° 3', land was discovered to the
eastward. The gulf or strait between the two lands was found to
be 25 miles wide, with 90 fathoms in mid-channel.

At 2 a.m. on the 29th we came in sight of islands, and on landing found a boat and depôt of provisions which had been deposited there the previous year by the *Investigator*. The strait is
here 4 or 5 leagues wide, with a depth of 50 and 60 fathoms in mid-
channel.

At 2 p.m. ice was seen on either shore of the channel. At midnight we worked up to the edge of the pack and could see round
both points, but further progress was blocked by floes of ice resting
on both shores. Our furthest point reached in that direction was
lat. 73° 30'[2], and long. 114° 35, and to the eastward, 73° 25', and
114° 14'. The ice was found to be streaming in on both sides of
the strait. In returning to the southward, the current which had
aided us in our progress northerly through the straits at an average
of 2 knots per hour, now assisted our return, and is therefore caused
by the wind.

On September 3rd Nelson Head was reached; the cliffs here rise
very abruptly from the sea to the height of 800 feet, being streaked
red horizontally, which on landing was found to be occasioned by
iron ore. At 2½ miles from the shore a depth of 117 fathoms was
found.

On the 7th the packed ice extended from n. by w. to w.s.w., and
the open water between it and the land so strewed with floes as to
render navigation difficult. A cairn was erected on an islet in lat.
72° 52', and long. 125° 24', and we returned to the south, searching
the coast as we went along for any harbour fit to winter in without
success, until we reached the entrance of Prince of Wales Strait,
where a secure position was found in Walker's Bay in lat. 71° 35',
long. 117° 35', on September 15th. Bay ice made the first week in
October, but the ship was not finally frozen-in until the 21st. The

---

[1] The *Investigator* was here on the 14th.
[2] This position is 57 miles from the furthest western point reached by the
Heela, and is the nearest approach to the accomplishment of the n.w. passage by
ships.

Eskimo left us in November and returned on May 25th. In the sledge travelling along the coast of Prince Albert Land drift-wood was fallen in with in small quantities until Peel Point was reached. On this point the ice was piled 30 feet high. The beach, which had hitherto been gravel, now became mud intermixed with sharp stones, and was upturned by the pressure of the ice.

The *Resolution* sleigh, Lieutenant Parkes, started from the head of Prince of Wales Sound on May 7th for Melville Island, and upon the following day got among hummocks that rendered travelling with the sleigh very difficult; on the 9th, not being able to find a passage for the sleigh, it was left behind in lat. 73° 31'. Melville Island was sighted on May 12th, and they landed under Cape Providence on May 16th. Lieutenant Parkes travelled along the coast towards Cape Hearne, coming across sleigh-tracks which we now know to have been those of Captain M'Clure, who passed along here a fortnight previous.

On the 17th they left Melville Island, and reached the tent on the 21st.

The ship moved in the ice on July 19th, but was not able to leave Winter Cove until August 5th; and in consequence of the ice resting on both shores we did not lose sight of our winter-quarters until the 30th. After running up to the head of Prince Albert Sound, and proving it to be a gulf and not a strait, on September 12th, the Dolphin and Union Strait was entered on the 17th.

On August 29th, 1853, the *Enterprise* (having left Cambridge Bay on the 9th) arrived at Cape Bathurst.[1] In passing the entrance of the Mackenzie, a much larger quantity of ice was observed than had been met with in 1851. On the 2nd it was calm, and an easterly set of 1·2 knots per hour was observed. The Pelly Islands were passed on the 3rd, and Herschel Island on the 5th of September. Here we found our progress to the westward barred by a close pack resting on the shore. On the 8th, the wind changing to the north-east, caused the ice to slacken, and when the fog cleared off we found we had been driven back to Point Kay, 40 miles to the eastward, since midnight of the 5th. After blasting a passage through the pack with gunpowder, we succeeded in reaching Herschel Island a second time on the evening of the 9th. The ice resting on the shore caused great delay, and we did not pass Flaxman Island until the 15th, and made fast to a grounded floe in 7½ fathoms in Camden Bay, lat. 70° 5', long. 144° 50', on the 16th, the easterly wind having packed the ice close on Brownlow

<hr>

[1] For the voyage of the *Enterprise* through the Dolphin and Union Strait, see page 153.

Point. On the 26th, young ice began to make, and on the 29th it was 2 inches thick, and, owing to pressure, cracked. On October 3rd, the land-water being completely frozen over, sleighs left the ship, and found abundance of drift-wood on the beach.

On May 21st, 1854, pools of water began to make on the flow, and on June 19th the communication with the shore was cut off, except by boat. On July 1st, a large party of Barter Island Eskimo, forty-one in number, came off in their kayaks, from whom a paper, printed on board the *Plover* at Point Barrow, was obtained, by which we learnt that the *Investigator* had not been heard of. The ice being sufficiently open alongshore on the 10th, the whale-boat under the command of Lieutenant Jago was despatched to Point Barrow to communicate with the *Plover*, and instruct Captain Maguire to obtain supplies sufficient to enable the *Enterprise* to return to the eastward to look after our consort.

The whale-boat was obliged to be launched across the ice frequently, so much so that on her arrival at Point Barrow her garboard streaks were nearly worn through. She arrived at Point Anxiety, July 12th; Point Milne, July 15th; Point Tangent on the 22nd, and at Point Barrow on the 24th. The *Plover* had left on the 20th. On July 30th, a sail was seen about 5 miles to the south-west, which afterwards proved to be H.M.S. *Rattlesnake*.

The ice broke up at the ship on July 15th, and enabled her to be moved as far as Point Brownlow, but the ice prevented farther progress, and she was driven back to her winter-quarters on the 18th by a westerly wind. This was, so far, fortunate, as it enabled the Barter Island Eskimo to bring the Rat Indians on board, the Chief of whom produced a paper, on which was written as follows :—

"Fort Youcon, June 27th, 1854.

"The printed slips of paper delivered by the officers of H.M.S. *Plover* on the 25th of April, 1854, to the Rat Indians were received on the 27th of June, 1854, at the Hudson Bay Company's establishment, Fort Youcon. The Rat Indians are in the habit of making periodical trading excursions to the Esquimaux along the coast. They are a harmless, inoffensive set of Indians, ever ready and willing to render every assistance they can to the whites.

"WM. LUCAS HARDISTY,
"Clerk in Charge."

The ice prevented the ship making much progress, and it was the 26th before Return Reef was reached. At noon, on the 29th, the Point Barrow natives met us. On the 6th, Harrison Bay was reached, and the ship arrived at Point Barrow on the 8th of August.

EXPLORATION of the COAST between the MACKENZIE and the
BACK RIVERS.

Journey of Samuel Hearne to the Northern Ocean in 1769-70-71-72.—
After two attempts to get to the northward, in the first of which
the guides failed him, and in the second he had the misfortune to
break his quadrant and to be plundered by the Indians, Mr. Hearne
set out for a third time on December 7th, 1771, with an Indian,
named Matonabba, as his guide, and on April 8th arrived at a river,
called by the natives Thelewey-aza-yeth. Here they collected bark
and wood for the canoes. On May 3rd they arrived at Clowey
Lake, where the canoes were built, and a large number of Indians
joined the party to make war on the Eskimo, which Hearne
endeavoured to dissuade them from. On July 14th, 1771, the
Coppermine River was reached. On the 17th the Eskimo were
fallen in with, and being surprised at night, were put to death
unmercifully, notwithstanding all Mr. Hearne's endeavour to check
the carnage. On the 18th the mouth of the river was reached.
After leaving the Coppermine River, a route further to the west
was taken in order to obtain provisions. On September 3rd they
arrived at Point Lake, where they camped in the neighbourhood
of Scrubby Wood. Continuing their course to the south-west by
slow marches, Athapusco Lake was reached on December 24th.
After expending some days in hunting beaver and deer, the lake
was crossed on January 9th, 1772, and Lake Clowey, where the
canoes were built, on the 15th of February; and upon June 30th
they returned to Prince of Wales Fort, having been absent 18
months and 23 days on this last expedition.

Captain Franklin's First Journey to the Shores of the Polar Sea in
1819-20-21-22.—Captain Franklin, accompanied by Dr. Richardson,
Mr. George Back, and Mr. Robert Hood, embarked on board Hudson
Bay ship *Prince of Wales*, on May 23rd, 1819, and arrived at York
Factory on August 30th. Taking their departure on September
9th, Norway Point was reached on October 6th, and Cumberland
House on the 23rd, where they passed the winter.

On January 19, 1820, Captain Franklin and Mr. Back proceeded
to the northward on two carioles and two sledges, drawn by dogs,
and arrived at Fort Carlton on February 1st. Leaving on the 8th,
on the 16th they reached Fort MacFarlane, which they left again on
the 20th, and got to Hudson Bay House on the 23rd. Starting again
on March 5th, N. W. Company's House was visited on the 9th, and

Pierre au Calumet on the 19th, and Fort Chipewyan on the 27th; having accomplished the following distances in miles :—

	Miles.
Cumberland House to Carlton House ..	263
Carlton House to Isle à la Crosse	230
Isle à la Crosse to Methye Portage	124
Methye Portage to Fort Chipewyan	240
Total	857

Here a canoe was built for the expedition—length, 32 feet 6 inches; extreme breadth, 4 feet 10 inches; depth, 1 foot 11 inches; 73 hoops of thin cedar, and will carry about 3300 lbs. weight. The weight of the canoe is about 300 lbs. On July 13, Dr. Richardson and Mr. Hood arrived.

Leaving Port Chipewyan in three canoes, containing five officers, one seaman, eighteen Canadians, and three interpreters, on July 18, 1820, after several portages, Moose Deer Island was reached on the 24th, and Fort Providence on the 28th. Leaving it on the 2nd, they arrived at their winter-quarters, Fort Enterprise, in lat. 64° 28', long. 113° 6', on August 19th. The length of the portages traversed was 21½ miles, and the total length of the voyage from Chipewyan 553 miles.

On September 9th, Sir John, accompanied by Dr. Richardson, set out on a pedestrian journey to the Coppermine River, which was reached on the 12th, the distance travelled to and fro being 110 miles. Mr. Back, in the meantime, went to Fort Chipewyan, and returned, performing the journey (upwards of 1000 miles) on foot.

On June 14th, 1821, the expedition left Fort Enterprise, and reached the head waters of the Coppermine on the 28th. Pursuing their journey, partly on the water and partly on the ice, they embarked, finally, on the 2nd of July, and met the Eskimo on the 15th, and encamped at the Bloody Falls on the 17th. Here Mr. Wentzel left them; and the remainder of the party, consisting of twenty persons, proceeded to sea. The distance hitherto travelled over was 334 miles, of which the canoes and baggage were dragged over snow and ice for 117 miles. The Coppermine River brings down no drift-wood. Berens Isle was reached on the 21st, where small drift-wood was found. 24th. "During the last two days the water rose and fell about 9 inches: the tides, however, were very irregular, and we could not determine the direction of the ebb or flood. A current set to the eastward, 2 miles per hour, during our stay." Point Barrow was rounded on the 26th. Arriving at Back River, shoals of capelin were seen, and small pieces of willow, which enabled them to make a fire. At Bathurst Inlet, on August 3 and 4, a fall of more than 2 feet water during the night was observed. Melville Sound was discovered on the 12th. Here the canoes were found to be much damaged by the heavy seas they had been exposed to. Point Turnagain was reached on the 21st, having traced 555 miles of coast-line since leaving the Coppermine.

Setting out on their return on the 22nd, Hood River was gained on the 25th; and here it was determined to abandon the canoes and cross the Barren Grounds. Obtaining a deer now and then, but feeding chiefly on *tripe de roche*, after undergoing great privation, the Coppermine River was reached on the 26th, and Mr. Back sent forward, who returned to them on October 1st; reporting barren country on this side, it was determined to make an effort to cross the river, which was done with great difficulty on the 4th in a coracle made by Mr. Back out of an old painted cover and willows, when Mr. Back was directed to go to Fort Enterprise. On the 6th, Mr. Hood being very weak, Dr. Richardson, with Hepburn, proposed to remain by him, while Sir John and the remainder of the party were to endeavour to reach Fort Enterprise; but on reaching it it was found to be perfectly desolate. A note from Mr. Back stated he had gone in search of succour. Feeding on deerskins, old bones, and *tripe de roche*, they passed a terrible existence, and were joined by Dr. Richardson and Hepburn on the 29th. Dr. Richardson then acquainted Sir John with the fate of poor Hood, and the necessity he was under of putting Michel to death. At length, on November 7th, relief, dispatched by Mr. Back, reached the party The Fort was left on the 12th, and Fort Providence reached on December 11th, and Moose Deer Island on the 17th.

Dr. Richardson and Mr. Kendall in the two Boats, 'Dolphin' and 'Union,' from the Mackenzie to the Coppermine Rivers.—The instructions received were to trace the coast between the Mackenzie and the Coppermine Rivers, and to return from the latter overland to Great Bear Lake. Leaving Fort Franklin on the 4th of July, 1826, Richards Island was reached on the 7th, Refuge Cove on the 8th, Cape Dalhousie on the 15th, Cape Bathurst on the 18th; a strong flood-tide setting to the westward; several whales seen. Franklin Bay was crossed on the 22nd, Cape Lyon on the 25th, when they were detained two days by a gale of wind. The tides were found to be regular, and the rise and fall 20 inches. Point De Witt Clinton was reached on the 29th, where they were stopped by the closeness of the ice. On August 4th land was discovered to the north, to which the name of Wollaston was given; and to the straits the

name of the two boats, *Dolphin* and *Union.* Near Manners Sutton
Island the tide indicated a stronger current of both flood and ebb
than we had hitherto seen; sometimes it attained a velocity of
3 knots per hour. Cape Krusentern was reached on the 7th, and
the mouth of the Coppermine River on the 9th. The boats were
abandoned at the Bloody Falls. The loads amounted to 72 lbs. per
man, and the pace averaged 2 miles per hour. On the 13th the
banks of the river were left, and a direct course made for the Great
Bear Lake, which was reached on the 17th. Indians were met with
on the 15th. On the 24th Benlim arrived in a boat and several
canoes from Fort Franklin, which they reached on September 1st.

TABLE of HIGH WATER reduced to full and change, compiled by LIEUT. KENDALL, R.N.,
on the BOAT VOYAGE between the MACKENZIE and the COPPERMINE RIVERS in 1826.

Date.	Name of Place.	Lati-tude.	Longi-tude.	Time of High Water reduced to full and change.	Wind, Direc-tion, and Force.	Remarks.
1825. Aug. 16	Garry Island	69°29	135°41	10·19	N.E. 6.	No ice in sight.
1826. July 9	Point Toker	69·38	132·18	1·45	N.E. by E. 5.	Rise 29 inches.
,, 10	,, ... }	69·43	131·58 {	0·56	E. 8.	Heavy ice.
,, 12				1·48	E. 5.	Little ice.
,, 13	Atkinson Island ..	69·55	130·43	0·32	S.E. 1.	Rise 18 inches.
,, 14	Boswell Cove	70·00	130·20	1·12	W. 6.	{Very little rise and fall.
,, 18	Point Sir P. Maitland	70·08	127·15	3·47	Calm.	In Harrowby Bay.
,, 19	Near Cape Bathurst..	70·33	127·21	1·28	E.S.E. 6.	Flood from eastward.
,, 20	Point Fetton	70·11	126·14	3·18	N.W. 6.	Rise 18 inches.
,, ,,	W. Horton River ..	69·50	125·55	3·15	W.N.W. 9.	
,, 21	,, ,, ...	..	..	3·49	W.N.W. 7.	
,, 27	Cape Lyon	69·46	122·51	6·33	E.N.E. 8.	Flood from eastward.
,, 30	{3 miles from Buchan River .. .. }	69·24	120·03	8·20	N.N.W. 8.	Rise and fall 9 inches.
Aug. 1	Point Wise..	69·03	119·00	7·04	W. 4.	Compact ice.
,, 3	Stapylton Bay	68·52	116·03	8·22	E. 2.	Bay filled with ice.
,, 4	{Between Cape Hope and Cape Bexley ..}	68·57	115·48	8·25	E.S.E. 4.	
,, 5	Chantry Island.. ..	68·45	114·23	7·22	W.S.W. 3.	
,, 6	{Seven miles from Cape Krusenstern .. ..}	68·32	113·53	7·13	Variable.	{Flood from S.E. Velocity 3 miles.

DISTANCES TRAVELLED BY DR. RICHARDSON AND MR. KENDALL.

		MILES.
From Fort Franklin to Point Separation		525
„ Point Separation to Point Encounter		159
„ Point Encounter to Coppermine River		863
„ Coppermine River to Fort Franklin		433

Sir G. Back's Voyage down the Great Fish River.—In the year
1832 grave apprehensions arose for the fate of Sir J. Ross and
his companions, who had left England in 1829. Sir George Back,
than whom no person was better qualified, undertook to command
an expedition down the Great River Thlew-ee-chow-dezeth. This

river, hitherto unvisited by any European, Sir George had become in some measure acquainted with by the accounts of the
Indians; and from their report it exceeded the Coppermine
both in extent and volume. As it was known Sir John Ross had
determined to effect the North-West Passage by Prince Regent
Inlet, the Thlew-ee-chow-dezeth (which has now received appro-
priately the name of Back) was thought to be the best route for
affording assistance to the missing expedition.

Accompanied by Dr. King and three men, Sir G. Back left
England on February 7, 1833, and passing through the United
States and Canada, they reached Fort William, on Lake Superior,
on May 20th, and left Norway House on June 28th, and arrived at
Fort Resolution on the Great Slave Lake on August 8th. Passing
through Artillery, Clinton Colden, and Aylmer Lakes by a short
portages, the river which was to conduct them to the Arctic Ocean
was gained; but the season was too far advanced to admit of their
reaching the Polar Sea this season ; the farthest point reached was
found to be in lat. 64° 41', long. 108° 8', and they returned to Fort
Reliance on September 7th.

Second Voyage.—Leaving Fort Reliance on the 7th of June, 1833,
they reached the boats, which had been built on Artillery Lake, on
the 10th, Lake Aylmer on the 24th, and the portage on the 28th;
and at 1 P.M. on the same day the boat was launched on the Back
River, which was still encumbered with ice. On the 4th of July
Mr. McLeod, who had hitherto accompanied them with a hunting
party, left ; and on July 8th, the ice having broken up, the boat
was launched on the river. Lake Beechey was reached on the 15th,
Lake Garry on the 21st, Lake Franklin on the 28th, at the northern
end of which they met the Eskimo. On the following day the
mouth of the river was reached. Arriving at Montreal Island on
August 2nd, a rise and fall of tide was found amounting to
12 inches, high-water being at 11.40 A.M. Parties were dispatched
in all directions to see if there was any possibility of creeping
alongshore among the grounded pieces of ice, but without success.
On the 5th the ice moved off a little, and enabled them to launch
the boat; Point Duncan was reached on the 6th, by watching their
opportunity ; Point Ogle on the 10th; here a log of wood, 9 feet
long and 9 inches diameter, was found, which was considered undoubted proof of the sea being open to the westward, and that the
main line of the land had been reached, in fact, Point Turnagain,
which had been reached by Franklin on August 21st, was only 4
miles north of this position. Setting out on their return on

the 16th, ascending the long and dangerous line of rapids, Lake
Garry was reached on August 31st. Traces of Eskimo were
found as high as Baillies River. On September 17th Mr. McLeod
was met with near Icy River, crossing the portage to Lake Aylmer:
the boat was navigated through Clinton Colden and Artillery
Lakes as far as Anderson's Fall, where it was left on the 25th;
and crossing over the mountains, Fort Reliance was reached ou
September 27th.

*Voyage of Messrs. Dease and Simpson from the Coppermine to the
Great Fish River in* 1838.—Leaving Fort Confidence at the north-east
end of the Great Bear Lake on June 7th, the ascent of the Dease
River was begun. On reaching its summit the boats were placed
on stout iron-shod sledges, and by dint of sailing and dragging
they were propelled across the Dismal Lakes on the ice, and were
launched on the Kendall River on the 19th. Waiting the disruption of the ice, the Coppermine River was gained on the 22nd,
the floods rendering the navigation very hazardous, and they were
arrested about a mile above the Bloody Fall on the 26th by the
ice. After a halt of five days, the Fall was descended on July
1st, the portage occupying six or seven hours; the boats had to be
carried half a mile. On July 2nd they met the Eskimo. Detained by
the close condition of the ice until the 17th, they obtained by their
nets 140 fish. Leaving the mouth of the Coppermine on that day,
they had great difficulty in forcing their way through the ice, and
did not reach Point Barrow until the 29th, and even then new ice
of considerable thickness formed during the night. The tides and
currents are very irregular, depending on the wind and ice, but
on no occasion was a change of more than 1 foot in the level
noticed. Cape Flinders was reached on August 9th: here they
were detained ten days by violent gales from the north and west,
in lat. 68° 16', long. 109° 21', Mr. Simpson proceeding to the eastward on foot with five of the company's servants and two Indians,
each man carrying half a cwt. Reaching Cape Alexander on the
23rd, he found an open sea to the east, and discovered land to the
north, to which he gave the name of Victoria. Returning westerly.
Boathaven was reached on the 29th. A furious gale from the west
detained them until the 31st; but they were enabled to regain
the Coppermine River on September 3rd. The boats were passed
up the Bloody Falls on the 5th with some damage. Nothing but
the skill and dexterity of the guides long practised like ours in
all the intricacies of river navigation could have overcome so many
obstacles. The boats were deposited 6 miles below the junction of

the Kendall with the Coppermine on the 10th. Striking straight out for the Kendall River they came upon it half a league below their Spring Provision Station. On the 12th the Hare Indians were met with, and on the 14th Fort Confidence was reached.

Second Journey.—Leaving Fort Confidence on June 15th, 1839, the Kendall River was reached on the 19th, and they learnt that the ice had cleared out of the Coppermine River ten days earlier than last year. On the 22nd the Bloody Fall was run in eleven hours ; but the sea-ice was still solid. Leaving the mouth of the Coppermine on July 3rd, they did not reach Cape Barrow until the 18th ; and to their great delight found Coronation Gulf open, and reached Boat-haven on the 20th, and Cape Alexander on the 26th, where a rapid tideway was experienced. It was high-water at noon. Full moon, the flood came from the westward, and did not exceed 2 feet. The temperature of the water 4 feet below the surface was 35°, and the air 56°. By attending to the tide Trap Cape was rounded; and on the last day in July a river was discovered, which was named the Ellice, and which is much larger than the Coppermine, and here no drift-wood comes down. Detained by the ice until the 5th of August, Point Seaforth was gained on the 11th, and upon the 13th they reached Sir George Back's Point, Sir C. Ogle thus connecting the Coppermine with Back River. On the 16th Montreal Island was visited. Having thus completed their instructions, these enter-prising men, taking advantage of the open season, crossed over to the land seen to the eastward, and reached their farthest point in this direction on the 19th, in lat. 68° 28', and long. 94° 14'. Cross-ing over on the 24th to what they conjectured to be part of Boothia, but which now proves to be King William Island the coast was traced for nearly 60 miles, until it turned up north, in lat. 68° 41', long. 98° 22', only 57 miles from Sir James Ross's Pillar. This cape was named Herschel ; and as the remains of one of the crew of either the *Erebus* or *Terror* was found by Sir L. McClintock to the southward and eastward of this cape, the first discovery of the North-West Passage, that is to say, a continuous sea from the Atlantic to the Pacific, rests with Messrs. Dease and Simpson, and with the expedition under Sir John Franklin.

Keeping to the northward, they crossed the Victoria Straits and reached Cape Colborne on the 6th, and coasting along the shore, discovered two bays, to which the names of Cambridge and Welling-ton were given : they crossed over to the southern shore on the 10th, and reached Wentzel River, where drift-wood was found, and on the 16th of September reached the entrance of the Coppermine, after the longest voyage ever performed in boats on the Polar Sea, viz., 1408 geographical miles. The boats were left at the Bloody Falls, and the land journey to the Great Bear Lake commenced. On the 24th the Dease River was reached, where they found a boat awaiting them, and Fort Confidence was reached the same afternoon.

Sir John Richardson and Dr. Rae's Voyage down the Mackenzie and along the Coast to the Coppermine River in 1848.—Leaving Liverpool on March 25th, they reached New York in a fortnight, and proceeded to Montreal ; and from thence to the Sault St. Marie, where they were detained some days, awaiting the breaking up of the ice on Lake Superior. Cumberland House was reached on June 15th, and the Mackenzie on the 15th of July. The sea was reached on the 4th of August. On the 22nd they were detained by the ice at Point Cockburn ; and it was only at the end of the month that they reached a bay between Capes Hearne and Kendall, where the boats were abandoned. Setting out on foot on September 3rd, and upon the 13th day reached Fort Confidence.

Dr. Rae's Journey to the Coast in 1851.—Leaving Fort Confidence on April 25th, with two men, on May 1st he reached the Polar Sea, near the mouth of the Coppermine. On the 4th they gained Port Lockyer, where they found some wood for cooking. On the 9th they reached lat. 68° 38′, and long. 110° 2′. Returning to the west, Douglas Island was gained on the 15th, and drift-wood found. Crossing over to Wollaston Land on the 16th, Eskimo were fallen in with near Cape Hamilton ; they had abundance of seals' flesh. On the 22nd, lat. 70° 0′ and long. 117° 17′ was gained, and called Cape Baring. Returning to the eastward on the 24th, on the 30th the Dolphin and Union Strait was crossed to Cape Krusentern in as direct a line as the rough ice would admit. On June 4th Richardson Bay was reached. The consumption of food in 33 days was 54 lbs. of flour, 128 lbs. of pemmican, 1¼ lb. of tea, 2 lbs. of chocolate, and 10 lbs. of sugar : no tent was carried. Leaving the coast on the 5th, the Kendall was reached on the 10th. The total distance travelled over from Fort Confidence is 942 miles.[1]

Second Journey.—On June 13th, three days after his arrival, the boats joined him at the Kendall River from Fort Confidence, having occupied 6½ days in the voyage. On the 15th the Coppermine was reached ; but the ice did not clear away until the 28th, and the sea

---

[1] On recomputing the distance, I make it 1100 miles, or about 25 miles per day, including three days' detention.—J. R.

was reached on July 5th. Point Barrow was rounded on July 16th, and Cape Alexander on the 24th. The ice breaking up on the 27th, the strait was crossed to the Finlayson Islands on the 27th, and Cape Colborne reached on August 1st.

At Parker Bay the flood tide came from the eastward. Reaching the south end of Taylor Island, they found very heavy, closely-packed ice; but the ebb tide being in their favour, they made way, but with considerable risk.

On the 6th Cape Princess Royal was discovered, and some drift-wood (poplar) was seen. In lat. 69° 56′, long. 102° 31′, a piece of pine, 18 feet long by 10 inches diameter, was found. On the 9th of August the ice was found close in to the shore.

After waiting until the 12th without the ice opening, Dr. Rae started on a foot journey, and eventually reached lat. 70° 3′, long. 101° 25″. Returning, the boats were reached in 8½ hours. The distance of the boat from the position of the *Erebus* and *Terror*, where they were abandoned on April 12th, 1848, is only 50 miles, being the nearest approach to the accomplishment of the North-West Passage by sea.[2] After attempting to cross over to King William Land, he set out on his return on the 16th.

Parker Bay was reached on the 20th, and Eskimo met with; here a piece of pine-wood, 5 feet 9 inches long, and round, resembling the butt end of a small flagstaff, was found; a bit of white line was nailed on to it with two copper tacks; both line and tacks had the Government mark. On the 22nd Point Back was gained, and on the 28th the Bloody Falls were reached, not having seen a bit of ice since leaving Point Back: 21 deer had been shot on the coast. Leaving one boat behind, the rapids were passed with great difficulty, and the Kendall River reached on the 5th day, and Fort Confidence was reached in the boat on September 10th.[3]

Voyage of the 'Enterprise' from Winter Cove, through the Dolphin and Union Strait, to Cambridge Bay.—The thickness of the ice in our winter-quarters, in lat. 71° 36′ and long. 117° 40′, attained its maximum, 5 feet 7½ inches, on April 1st, 1852. On May 1st it was 5 feet 3 inches; on June 1st, 5 feet 1 inch; on May 1st, 4 feet 10 inches. The ship forged ahead in her icy cradle on July 16th, the thickness of the ice then being 3 feet 4½ inches, and on the 19th

[1] Two of Dr. Rae's men reached 70° 13′, and saw coast 7° further.
[2] The nearest approach of two ships is the *Hecla* and *Enterprise*, 57 miles.
[3] On the coast of Victoria Land the flood-tide comes from the east to long. 104° or 105°, where it is met with the flood coming from N.E. down the Victoria Channel.—J. R.

she swung to the wind; on the 28th the temperature of sea at
surface was 34·5°, but it was not until August 5th the ship was
able to proceed to sea. The ice resting on the shore on either side,
detained her within sight of Winter ·Harbour until the 8th of
September, during which time whales were seen and seals were
numerous. Entering Prince Albert Sound, it was on the 13th
found to be a gulf and not a strait. Having now discovered that
Wollaston, Victoria, and Prince Albert Land are all one, it was
determined to enter the Dolphin and Union Strait, which was done
on September 17th. Sutton and Liston Islands were reached on the
20th with very little obstruction from the ice, and on the following
day Cape Krusenstern was passed. On the 22nd, by a slant of wind,
72 miles were made; and when the ship was anchored the current
was found to set to the eastward, at one time as much as 1 knot
per hour. On the 23rd Cape Franklin was seen; and in the evening
we unfortunately got aground in Byron Bay. On the following
morning, on opening Wellington Bay, the wind freshened; and increasing to a gale, we ran back to the westward, where there was
more room, and underwent an equinoctial gale under close-reefed
topsails, with the thermometer at 11°. The sea froze as it lodged;
and it was late in the forenoon of the next day before the ice that
had made on board the ship during the night was cleared away.
Passing through the Finlayson group on the 26th, Cambridge Bay
was gained on the 27th; but the water shoaling suddenly, we
struck the ground, and remained fast until the ice set sufficiently
firm to allow of our removing everything out of the ship to the
shore; and the tides taking off, it was not until October 15th
that we got the ship afloat.

On crossing over to the Continent with a sleigh, in October, the
ice was found so rotten in the neighbourhood of Cape Trap that we
could not land. The mean temperature of the quarter ending
December 1852 was found to be 5° lower than that experienced
last year, though we were 2½° further south. The sleighs left the
ship on the 12th of April, and crossing over the Colborne Peninsula
came upon the sea-ice near Rae Inlet the following day. On the
23rd, in lat. 69° 10', long. 121° 20', came upon the junction of the
old and the new ice; the former being so hummocky as to be impracticable for sleighs. After exploring to the north, north-east, and
north-west, and finding nothing but a confused jumble of angular
pieces, some of which were upwards of 20 feet high, and between
which the snow was so loose that you frequently sunk up to your
middle, it was determined to strike in for the Victoria shore. By
unlading the sleighs, and carrying half-loads, Drift-wood Point (so

called from a small piece of much decayed wood being found on
it) was reached. This point is 30 miles from Cape Crozier on King
William Land, near which Sir L. M'Clintock found the boat. So
had we gone up the eastern instead of the western side of the
strait we should have discovered the relics. On the 8th of May
a cairn was reached, in which was contained a notice from Dr.
Rae, dated August 13th, 1851. Thus we learnt that our field of
search had been previously examined. On the 10th an island was
reached from which no land was visible, except in the direction we
had come from, and the appearance of the pack forbid all hope of
penetration even with a light load. During this portion of the
journey sludge ice and sometimes pools of water were found in the
neighbourhood of large hummocks, which at first I thought might
be caused by the increased weight of drifted snow causing the
hummock to break through the ice, but now I am of opinion that
it is occasioned by the set of the tide round these hummocks,
which are aground; the furthest point attained being in lat. 70° 35',
long. 101°. In returning to the ship several cracks in the ice
were seen, which were not there when we passed up. The ship
was reached on May 21st, after an absence of forty-nine days,
and the accomplishment of 753 miles, which does not include the
previous journeys laying out the depôts. In July the ice along the
shore began to melt, and large quantities of salmon were caught
by the seine. The result of our observations on the tides is as
follows. It is high-water on F. and C. days at 11.30, and the
rise and fall varied from 2 feet 4 inches to 7 inches. The set of
the tide was so irregular, and so dependent on the wind, that I
cannot say whether the flood comes from the east or the west.
On the 25th the ice began to move, but did not open sufficiently
to allow the ship to leave the bay until the 10th August; the
wind being light, we were driven by the current to the eastward,
and sighted Cape Colborne¹ the next day. Cape Alexander was
doubled at 1 A.M. on the 13th. At Douglas Island the ice was
found closely packed. On the 20th the ship was carried away in
the pack to the eastward at the rate of 1 mile per hour. Cape
Krusenstern was passed on the 23rd.

In the afternoon of the 27th, the wind drew round to the south-
west, and we made all sail out of the straits, but, the weather
being thick, ran close past Clerks Island without seeing it.
The wind drawing to the west, we were compelled to stand over
to Baring Land, and the next morning found ourselves off Darnley

¹ The easternmost position reached was in long. 105°, making 63¾° of longitude
sailed over after entering the Arctic circle.

Bay. Cape Parry was passed at midnight, and we came across some heavy ice, being the first met with since leaving the straits. On the 30th it was so close as to compel us to haul in shore, affording a great contrast with the state of the ice at the same period two years ago, when the pack was 30 miles from the land. Cape Bathurst was passed on September 1st. On the 2nd, the temperature of the sea rose to 36°, and several whales were seen. The current was found to set to the eastward, at the rate of 1·2 miles per hour, which so delayed our progress that Herschel Island was not passed until the 4th.

Dr. Rae's Journey from Repulse Bay across Rae Isthmus and Simpson Peninsula to the West Coast of Boothia Felix.—Passing the winter of 1853-54 on the head of Repulse Bay, where he maintained himself almost entirely by his own resources, on March 31st, he set off, accompanied by four men. Pelly Bay was reached on April 16th, and the Eskimo met with on the 20th, and on the 29th the mouth of the Murchison River: continuing his course along the shore of Boothia Felix, Cape Porter (so named by Sir John Ross) was reached on May 6th. After obtaining numerous articles from the Eskimo belonging to the *Erebus* and *Terror*, and receiving from them an account of the crews having perished by starvation, Dr. Rae returned to Repulse Bay, which was reached on May 26th. Leaving Repulse Bay on August 6th in the boats (the summer being extremely cold and backward), Churchill River was reached on August 28th, and York Factory on the 31st.

Voyage of Mr. J. Anderson down Back River.—Leaving Fort Resolution in three bark-canoes on June 22nd, 1855, on the 28th the ice was fallen in with at the Tal-thal-leh Lake; and it was not until July 2nd that the mountain was reached. Carrying everything across the portages, Lake Aylmer was gained on the 8th, and Sand Hill Bay on the 11th. Availing himself now of the information supplied by Sir G. Back, the river was descended, and notwithstanding the exquisite skill of our Iroquois bowmen, the canoes were repeatedly broken and much strained. On the 20th the Eskimo were met with below the Mackinlay River. On Lake Garry the ice still delayed their progress. On arriving at the rapids below Lake Franklin several articles belonging to the missing expedition were found among the Eskimo. On August 1st, Montreal Island was reached with considerable difficulty, and the remains of a boat and other things belonging to the ships were found. Crossing over to Elliot Bay on the 5th, the inlet was full of ice, and they could only proceed along shore at high-water.

The canoes were so leaky, that Mr. Anderson determined upon setting out on foot, and reached Maconochio Island on the 8th. " It was impossible to cross over to Point Richardson as I wished, the ice driving through the strait between it and Maconochio Island at a fearful rate." "No party could winter on this coast. In the first place, there is not enough fuel, and secondly, no deer pass." Returning up the river, Lake Aylmer was reached on the 31st, and Old Fort Reliance on September 11th.

Victoria Strait and Franklin Channel.—The record brought back by Sir Leopold M'Clintock informs us that H.M. ships *Erebus* and *Terror* wintered in the ice in lat. 70° 5′ N., and long. 98° 23′ W., and that the vessels reached this position in one season from Beechey Island: whether by Franklin or M'Clintock Channel is not known, but most probably by the former. The ships, it appears, were beset on September 12th, 1846, and during the following eighteen months were drifted only 12 miles to the south-west, when they were finally abandoned on April 12th, 1848. The following are a few extracts respecting the state of the ice in Franklin and Victoria Channel, from Sir Leopold M'Clintock's interesting journal. On August 21st, 1858, the *Fox* reached a position half-through Bellot Strait, which is scarcely one mile wide at its narrowest part. At the turn of the tide the vessel was carried back to the eastward at the rate of 6 miles per hour. "The tide runs through to the west from two hours before high-water to four hours after it: that is to say, the tide comes from the west, as is the case in Fury and Hecla Strait: the rise and fall is less on the west side than upon the east. On September 29th, the view from Cape Bird is thus described :—" There is now much water in the offing, only separated from us by the belt of islet-girt ice scarcely 4 miles in width." " The water runs parallel to the coast, and is 4 or 5 miles broad." On the 28th, the *Fox* was compelled, by the freezing of the ice, to take up her winter-quarters in Port Kennedy. Lieutenant Hobson, who had left the ship with sleighs on the 25th instant, returned to the ship on October 6th, having been stopped by the sea washing against the cliffs, in lat. 71½°. On the 19th, Lieutenant Hobson started again, and returned on November 6th. On the 25th, they camped on the ice ; a north-east gale sprang up, and, detaching the ice, blew them off shore, and they were not able to regain the land for two days.

The following records are made of the state of the ice in Bellot Strait during the winter:—October 7th.—" The weather is mild ; Bellot Strait is almost covered with ice, which drifts freely with

every tide." November 1st.—"Whenever we have a calm night
we can hear the crushing sound of the drift-ice in Bellot Strait,
which continues to open within 500 yards of the Fox Islands, and
emits dark chilling clouds of hateful, pestilent, and abominable
mist.

On February 17th, 1859, the sledge-parties started to carry out
the depôts. Advancing to the southward, the condition of the ice
is thus described :—" Throughout the whole distance we found a
mixture of heavy old ice and light ice of last autumn, in many
places squeezed up into the pack; but as we advanced southward
aged floes were less frequently seen." On March 1st the neighbourhood of the Magnetic Pole was reached, and the Eskimo seen.
The ship was reached on March 14th, having travelled 420 miles
in 25 days. Mr. Young and his party returned on board on March
3rd, having placed their depôt on the shore of Prince of Wales
Land, about 70 miles south-west of the ship, the shore of which
was found to be " fringed for a distance of 10 miles to seaward with
an ancient land-floe." The remaining width of the strait was about
15 miles, and this space was composed of ice formed since
September last. This was the water we looked at so anxiously
last autumn from Cape Bird and Pemmican Rock. On April 2nd
Sir Leopold and Lieut. Hobson started : the load for each man to
drag was 200 lbs., and for each dog 100. On April 20th, in lat.
70½° N., the Eskimo were met with. They had been as far north
as lat. 71¼° hunting seals. Crossing a wide bay upon level ice,
indicating much open water here late last autumn, the neighbourhood of the Magnetic Pole was reached on the 24th, and a detention
of three days on account of a heavy north-east gale was incurred.
At Cape Victoria, Lieut. Hobson parted company, going direct to
Cape Felix. Sir Leopold struck across this strait for Port Parry;
finding a rough pack it took him three days to traverse the strait.
Matty Island was reached on the 4th of May, and Point Booth on
May 10th, where a number of articles from the missing ships were
found. Crossing over to Point Ogle on May 12th, and Montreal
Island on the 15th. " Since our first landing on King William
Land we have not met with any heavy ice; all along its eastern and
southern shore, together with the estuary of this great river, is one
vast, unbroken sheet, formed in the early part of last winter
where no ice previously existed." Crossing over to the mainland,
near Point Duncan, on the 18th of May, they followed the coast
as far as Barrow Inlet, from whence they returned to King
William Land. On May 25th, a short distance to the east of Cape
Herschel, a skeleton was discovered, which, from the documents

and the clothing found on it, proved that one of the crew of the *Erebus* and *Terror* had certainly passed Cape Herschel, which had been previously reached from the westward by Dease and Simpson. Advancing along the west to the north, hummocks of unusually heavy ice were met with. On the coast from Point Victory northward the sea is not so shallow, and the ice comes close in ; to seaward all was heavy, close pack, consisting of all descriptions of ice, but for the most part old and heavy. Crossing over land to Port Parry, Sir Leopold reached his depôt there on June 4th, and Cape Victoria, on Boothia Felix, on the 8th, and reached the *Fox* on June 19th. With respect to a navigable North-West Passage, and to the probability of our having been able last season to make any considerable advance to the southward, had the barrier of ice across the western outlet of Bellot Strait permitted us to reach the open water beyond, Sir Leopold thus expresses himself :—" I think, judging from what I have since seen of the ice in Franklin Strait, that the chances were greatly in favour of our reaching Cape Herschel on the south side of King William Land, by passing, as I intended to do, eastward of that island. From Bellot Strait to Cape Victoria we found a mixture of old and new ice, showing the exact proportion of pack and of clear water at the setting in of winter. Once to the southward of the Tasmania Group, I think our chief difficulty would have been overcome, and south of Cape Victoria I doubt whether any further obstruction would have been experienced, as but little, if any ice remained. The natives told us the ice went away and left a clear sea every year." " No one who sees that portion of Victoria Strait which lies between King William Land and Victoria Land as we saw it, could doubt of there being but one way of getting a ship through it, that way being the extremely hazardous one of drifting through in the pack. The wide channel " (M'Clintock Channel) "between Prince of Wales and Victoria admits a vast and continuous stream of very heavy ocean-formed ice from the north-west, which presses upon the western face of King William Island, and chokes up Victoria Strait in the manner I have just described. I do not think the North-West Passage could ever be sailed through by passing westward, that is, to windward of King William Island." " Had Sir John Franklin known that a channel existed eastward of King William Land (so named by Sir John Ross), I do not think he would have risked the besetment of his ships in such very heavy ice to the westward of it; but had he attempted the North-West Passage by the eastern route, he would probably have carried his ships safely through to Behring Straits." " Perhaps some future voyager, profit-

ing by the experience so fearfully and fatally acquired by the
Franklin Expedition, and the observations of Rae, Collinson, and
myself, may succeed in carrying his ship through from sea to sea."
" In the meantime to Franklin must be assigned the earliest discovery of the North-West Passage, though not the actual accomplishment of it in his ships."

" The extent of coast-line explored by Captain Young (on Prince
of Wales Land) amounts to 380 miles, whilst that discovered by
Hobson and myself amounts to nearly 420 miles, making a total
of 800 geographical miles of new coast-line, which we have laid
down."

Lieut. Hobson, after parting with Sir Leopold at Cape Victoria,
thus describes the condition of the ice between Boothia Felix and
King William Island :—" No difficulty was experienced in crossing
James Ross Strait. The ice appeared to be of but one year's
growth, and although it was in many places much crushed up, we
easily found smooth leads through the line of hummocks. Many
very heavy masses of ice, evidently of foreign formation, have been
here arrested in their drift; so large are they that, in the gloomy
weather we experienced, they were often taken for islands." At
Cape Felix he observes :—" The pressure of the ice is severe, but
the ice itself is not remarkably heavy in character; the shoalness
of the coast keeps the line of pressure at a considerable distance
from the beach : to the northward of the island the ice, as far as I
could see, was very rough, and crushed up into large masses."

Having laid before you extracts from the journals of the different
expeditions which have reached the Arctic Sea from the Pacific
Ocean, the rivers of America, and that portion of Asia which is
in the immediate neighbourhood of Behring Straits, we are now in
a condition to comprehend fully the effect which the currents of
the Pacific have upon the motion of the ice to the north of Behring
Straits.

The first, and a very important point it is, which presents itself
is the contrast between the configuration of the two continents after
the narrow shallow strait has been passed which separates them. On
the western side the trend of the coast is gradual, affording immediate
access for the current to or from the strait along the shore of the
north face of the Asiatic continent. On the opposite side of the strait
the turn of the shore is abrupt and rectangular. On the Asiatic
side we have indisputable records of open water continuously met
with during the period of lowest temperature for a distance of
upwards of 1000 miles. On the opposite shore the ice is driven
frequently during the winter by the force of the wind from the

coast at Point Barrow, but along the American continent to the
eastward the ice, as far as we are capable of judging from one
winter's experience, it remains quiet and immovable. Hence comes
the question, Does the effect of the Pacific current lose itself in the
expanse of the Polar Sea, or does it take an easterly trend? So far
as experience guides us, the positions reached by the *Enterprise* in
1850 prove the existence of a loose pack 100 miles to the north-
east of Point Barrow; beyond this, until we come to the records
given by Sir R. M'Clure, nothing is known, but we have undoubted
testimony that the pressure on the north face of Banks Land comes
from the westward : and here in this strait, between Melville
Island and Banks Land, occurs one of those dead locks in the
motion of the ice that are remarkably instructive. We find the
Hecla prevented going to the westward along Melville Island by
the pressure of the ice on the land from the westward; and on the
opposite shore it became necessary to leave the *Investigator* to her
fate in Mercy Bay from the same cause. Though mention is
made in the first autumn of her incarceration of open water having
been seen along the coast to the eastward, yet in all the transits
across the straits on the ice in 1851, 1852, and 1853, we have no
record of any ice movement; whereas directly the channel east of
Melville Island is opened, the *Resolute* experiences an easterly drift.
I forbear to trespass upon the ground so ably and so laboriously
explored by the eastern expeditions, knowing that from some of
the officers engaged in the exploration from that side a much fuller
and more comprehensive account of the movement of the ice north
of the Parry Islands, and through Barrow Strait and Lancaster
Sound into Baffin Bay, can be given than it is possible for me to
do; but so far as can be gathered from the accounts given, it
may, I think, be assumed that the pack is looser, and open spaces
of water are more frequent to the north than they are to the south
of the Parry Group; and the effect of this current from the
Northern Sea, after checking the easterly set through M'Clure
Strait, assisted the passage of the *Erebus* and *Terror* from Barrow
Strait to King William Land. Though the Pacific current is in a
great measure turned aside from the face of the American continent by the abrupt change in the direction of the coast at Point
Barrow, the testimony of all navigators is conclusive that it is
felt, and that an easterly set pervades to a greater extent than a
westerly one, and that this set is more noticeable to the east of the
Mackenzie. The latter river, the Coppermine, the Ellice, and the
Back, no doubt contribute to the arrest of the pack in Victoria

M

Strait, and thus prevented the escape of the *Erebus* and *Terror* ; but it is more than probable that the detention of those two vessels in a position which differed only 12 miles in 18 months was mainly owing to the meeting of the currents which originally had but one origin, and that the Pacific.

ETHNOLOGY.

I.

PAPERS ON THE GREENLAND ESKIMOS.

BY

CLEMENTS R. MARKHAM.

1.

ON THE ORIGIN AND MIGRATIONS OF THE GREENLAND ESKIMOS.

AN expedition to the region round the North Pole will advance every branch of science, and will enrich the store of human knowledge generally. Its geographical discoveries will only be one out of the many valuable results that will be derived from it; but, as geographers, we may well look forward with deep interest to the rich harvest that will be reaped by our science, and take a preliminary survey of the additional knowledge that may be in store for us. It should be remembered that, though only one-half of the Arctic regions has been explored, yet that throughout its most desert wastes there are found abundant traces of former inhabitants where now all is a silent solitude. Those cheerless wilds have not been inhabited for centuries, yet they are covered with traces of the wanderers or sojourners of a by-gone age; and the unexplored region far to the north, even up to the very Pole itself, may not improbably be at this moment supporting a small and scattered population. The wanderings of these mysterious people, the scanty notices of their origin and migrations that are scattered through history, and the requirements of their existence, are all so many clues which, when carefully gathered together, will assuredly tend to throw some light on a most interesting subject. The migrations of man within the Arctic zone give rise to questions which are closely connected with the geography of the undiscovered portions of the Arctic regions—questions which can only be solved by a scientific Arctic expedition. The origin and history of the Eskimo of Greenland, and

N

especially of those interesting people on the northern shores of
Baffin's Bay, who were named by Sir John Ross the "Arctic Highlanders," are topics serving to illustrate one of the numerous points
which will engage the attention of the Arctic Expedition, and, at
the same time they may throw some passing light on questions
in Arctic physical geography which still remain unsolved.

Until within the last nine centuries the great continent of Greenland was, so far as our knowledge extends, untenanted by a single
human being—the bears and reindeer held undisputed possession.
There was a still more remote period when fine forests of exogenous
trees clothed the hill-sides of Disco, when groves waved, in a
milder climate, over Banks Island and Melville Island, and when
corals and sponges flourished in the now frozen waters of Barrow's
Strait. Of this period we know nothing; but it is at least certain
that when Erik the Red planted his little colony of hardy Norsemen at the mouth of one of the Greenland fiords, in the end of the
tenth century, he apparently found the land far more habitable
than it is to-day.

For three centuries and a half the Norman colonies of Greenland
continued to flourish; upwards of 300 small farms and villages
were built along the shores of the fiords from the island of Disco to
Cape Farewell[1] (for the persevering Danish explorer Graah has
truly conjectured and Mr. Major has clearly proved that the East
and West Bygds were both on the west coast),[2] and Greenland
became the see of a Bishop. The ancient Icelandic and Danish
accounts of these transactions are corroborated by the interesting
remains which may be seen in the Scandinavian museum at Copenhagen. During the whole of this period no indigenous race was
seen in that land, and no one appeared to dispute the possession of
Greenland with the Norman colony.[3] A curious account of a
voyage is extant, during which the Normans reached a latitude
north of Cape York; yet there is no mention of any signs of a
strange race. The Normans continued to be the sole tenants of
Greenland, at least until the middle of the fourteenth century.

But Thorwald, the boastful Viking, who sailed away west from
Greenland and discovered America,[4] did meet with a strange race
on the shores of Vinland and Markland, which probably correspond
with modern Labrador. Here he found men of short stature, whom
he contemptuously called Skrællings (chips or parings), and some
of whom he wantonly killed. Here, then, is the first mention of

[1] Egede.   [2] Graah's 'Greenland,' Introd. and p. 163.
[3] Crantz, i. p. 257.   [4] Ibid.

the Eskimo. At this period (the eleventh century) they had probably spread themselves from Northern Siberia, across Behring
Strait, along the whole coast of Arctic America, until they were
stopped by the waves of the Atlantic. The hostility of the Red
Indians was an effectual barrier to their seeking a more genial home
to the south. They were not likely to wander towards the barren
and inhospitable north any more than their descendants do to-day ;
and they had no inducement to trust themselves in their frail
kayaks, or *umiaks*, on the waves of the Atlantic. They assuredly
never crossed over to Greenland by navigating Davis Strait or
Baffin's Bay. This, as I believe, is the southern belt of Eskimo
migration ; but it is with the Greenland Eskimo that we have now
to do, who had had no communication with their southern brethren
since their ancestors hunted together on the frozen tundra of
Siberia, and who, after centuries of wanderings along wild Arctic
shores and in regions still unknown, first make their appearance in
Greenland, coming down from the north.

Our last historical glimpse of the Norsemen of Greenland shows
them living in two districts, in villages along the shores, with
small herds of cattle finding pasturage round their houses, with
outlying colonies on the opposite shores of America, and occasional
vessels trading with Iceland and Norway ; but no grain would
ripen in their fields. They seem to have been a wild turbulent
race of hardy pirates, and their history, short as it is, is filled with
accounts of bloody feuds. All at once, in the middle of the fourteenth century, a horde of Skrœllings, resembling the small men of
Vinland and Markland, appeared on the extreme northern frontier
of the Norman settlements of Greenland, at a place called Kindelfjord.[1] Eighteen Norsemen were killed in an encounter with
them ; the news of the invasion travelled south to the East Bygd ;
one Ivar Bardsen came to the rescue in 1349, and he found that all
the Norsemen of the West Bygd had disappeared, and that the
Skrœllings were in possession. Here the record abruptly ceases,
and we hear nothing more of Greenland until the time of the Eliza-
bethan navigators, and nothing authentic of either Norsemen or
Skrœllings until the mission of Hans Egede, in the middle of the
last century.

When the curtain rises again all traces of the Norsemen have
disappeared save a few Runic inscriptions, extending as far north

---

[1] Crantz, i. p. 258, quoting from La Peyrère, who repeats from Wormius.
After a careful consideration of the evidence, Mr. Major has concluded that
Kindelfjord is the inlet where the present Danish settlement of Omenak is
situated, north of the island of Disco.

as the present settlement of Upernivik, some ruins and the broken church-bells of Gardar. The Skrœllings or Eskimos, are in sole possession from Kingitok to Cape Farewell. And the ancient Norse records are fully corroborated by the traditions of the Eskimos, in the statement that they originally came from the north. Like the Mongolian races, the Eskimos are careful genealogists: Crantz tells us that they could trace back for ten generations;[1] and the story handed down from their forefathers is that they reached Southern Greenland by journeys from the head of Baffin's Bay.

The interesting question now arises—whence came these Greenland Eskimos, these *Innuit*, or men, as they call themselves. They are not descendants of the Skrœllings of the opposite American coast, as has already been seen. It is clear that they cannot have come from the eastward, over the ocean which intervenes between Lapland and Greenland, for no Eskimo traces have ever been found on Spitzbergen, Iceland, or Jan Mayen. We look at them and see at once that they have no, or only very remote, kinship with the red race of America; but a glance suffices to convince us of their relationship with the Tuski or northern tribes of Siberia. It is in Asia, then, that we must seek their origin, that cradle of so many races, and the search for some clue is not altogether without result.

During the centuries preceding the first reported appearance of Skrœllings in Greenland, and for some time previously, there was a great movement among the people of Central Asia. Tugrul Beg, Jingiz Khan, and other chiefs of less celebrity, led vast armies to the conquest of the whole earth, as they proudly boasted. The land of the Turk and the Mongol sent forth a mighty series of inundations which flooded the rest of Asia during several centuries, and the effects of which were felt from the plains of Silesia to the shores of the Yellow Sea, and from the valley of the Ganges to the frozen tundra of Siberia. The pressure caused by these invading waves on the tribes of Northern Siberia drove them still farther to the north. Year after year the intruding Tatars continued to press on. Shaibani Khan, a grandson of the mighty Jingiz, led fifteen thousand families into these northern wilds, and their descendants, the Iakhuts, pressed on until they are now found at the mouths of rivers falling into the Polar Ocean. But these regions were formerly inhabited by numerous tribes which were driven away still farther north, over the frozen sea. Wrangell has preserved traditions of their disappearance, and in them, I think, we may find a clue to the origin of the Greenland Eskimos.

---

[1] Crantz, i. p. 229.

The Iakhuts, it is said, were not the first inhabitants of the country along the banks of the river Kolyma.[1] The Omoki, a tribe of fishermen, the Chēlaki, a nomadic race possessing reindeer, the Tunguses, and the Iukahirs were their predecessors. These tribes have so wholly disappeared that even their names are hardly remembered. An obscure tradition tells how "there were once more hearths of the Omoki on the shores of the Kolyma than there are stars in an Arctic sky."[2] The Onkilon, too, once a numerous race of fishers on the shores of the Gulf of Anadyr, are now gone no man knows whither. Some centuries ago they are said to have occupied all the coast from Cape Chelagskoi to Behring Strait, and the remains of their huts of stone, earth, and bones of whales are still seen along the shores.[3] The Omoki are said to have departed from the banks of the Kolyma in two large divisions, with their reindeer, and to have gone northward over the Polar Sea.[4] Numerous traces of their *yourts* are to be seen near the mouth of the Indigirka. The Onkilon, too, fled away north, to the land whose mountains are said to be visible from Cape Jakan.

Here we probably have the commencement of the exodus of the Greenland Eskimo. It did not take place at one time, but spread over a period of one or two centuries. The age of Mongol invasion and conquest was doubtless the age of tribulation and flight for the tribes of Northern Siberia. The Khivan genealogist Abu-'l Ghází tells us that when Ogus Khan, a chief belonging to the conquering family of Zingiz, made an inroad into the south, some of his tribes could not follow him on account of the deep snow.[5] They were called in reproach *Karlik*, and this very word, in its plural form of *Karalit*, is the name which the Eskimos of Greenland give themselves; but I do not attach much weight to this coincidence.

The ruined *yourts* on Cape Chelagskoi mark the commencement of a long march; the same ruined *yourts* again appear on the shores of the Parry group—a wide space of 1140 miles intervenes, which is as yet entirely unknown. If my theory be correct, it should be occupied either by a continent or by a chain of islands; for I do not believe that the wanderers attempted any navigation, or indeed that they possessed canoes at all. They kept moving on in search of better hunting and fishing grounds along unknown shores, and across frozen straits, and the march from the capes of Siberia to Melville Island doubtless occupied more than one generation of wanderers.

---

[1] Wrangell, p. 171.   [2] Ibid., p. 53.   [3] Ibid., p. 358.
[4] Ibid., p. 181.   [5] Strahlenberg.

There is some evidence, both historical and geographical, that the unknown tract in question is occupied by land. A chief of the Tuski nation told Wrangell that from the cliffs between Cape Chelagskoi and Cape North, on a clear summer day, snow-covered mountains might be descried at a great distance to the north.[1] He maintained that this distant northern land was inhabited, and added that herds of reindeer had been seen to come across the frozen sea, and return again to the north. The Tuskis also spoke of a much more northern land, the lofty mountains of which were visible on very clear days from Cape Jakan.[2] Wrangell himself never saw this mysterious land, and the Tuskis were hardly believed until it was actually re-discovered by Captain Kellett, in the *Herald*, in 1850. In August of that year he sighted an extensive and high land to the north and north-west of Behring Strait, with very lofty peaks, which is believed to be a continuation of the range of mountains seen by the natives off Cape Jakan.[3] There are geographical reasons, which have been pointed out by Admiral Sherard Osborn, for the supposition that land, either as a continent or as a chain of islands, extends to the neighbourhood of the westernmost of the Parry group. The nature of the ice-floes between the north coast of America, off the mouths of the Colville and Mackenzie, and Banks Island, leads to the conclusion that the sea in which such ice is formed must be, with the exception of some narrow straits, land-locked. The Eskimos of this part of the coast of North America are never able to advance more than 30 miles to seaward.[4] The ice is aground in 7 fathoms of water, and the floes, even at the outer edge, which are of course lighter than the rest, are 35 to 40 feet thick. The nature of the ice is the same along the west coast of Banks Island. When the *Investigator* made her perilous voyage along this coast, the channel between the ice and the cliffs was so narrow that her quarter-boats had to be topped up to prevent their touching the lofty ice on one side and the cliffs on the other. The pack drew 40 or 50 feet of water; it rose in rolling hills upon the surface, some of which were 100 feet high from base to summit, and when it was forced against the cliffs it rose at once to a level with the *Investigator's* fore yard-arm.[5] McClintock also mentions the very heavy polar ice which is pressed up on the north-western shore of Prince Patrick Island.[6]

Such awful ice as this was never seen before in the Arctic regions. The only way of accounting for its formation, which must have

[1] Wrangell, p. 326.   [2] Ibid., p. 342.
[3] Osborn's 'North-West Passage,' p. 49.   [4] Ibid., p. 70.
[5] Ibid., p. 201.   [6] 'Blue Book,' p. 569. (Further papers, 1855.)

taken a long course of years, is that it has no sufficient outlet, and
that it goes on accumulating from year to year. It must therefore
be in a virtually land-locked sea, and this of course implies land to
the north, as well as to the east, south, and west. Captain Cook
supposed there must be land to the north, from having observed
great flocks of ducks and geese flying south in September. Dr.
Simpson tells us that the natives of Point Barrow have a tradition
that there is land far away to the northward, and that some of
their people once reached it. It was a hilly country, inhabited by
men like themselves, and called *Iglun-nuna.*[1] Here, then, is my
bridge by which the Omoki, Tunguses, and Onkilon passed over
from the frozen tundra of Siberia to the no less inhospitable
shores of Prince Patrick's Island, to those of the head of Wellington
Channel and Baffin's Bay, and far into the unknown region. The
theory of Eskimo migration is thus illustrated by facts in physical
geography.

On Melville and Banks Islands, and near Northumberland Sound,
we meet with the same ruined *yourts* of stone and earth, the same
stone fox-traps, and the same bones of whales and other animals as
were seen by Wrangell at the mouth of the Indigirka. These traces
were met with by the Arctic expeditions all along the shores of the
Parry group, from Prince Patrick's Island to Lancaster Sound, a
distance of 540 miles. They were of great antiquity, and had evidently not been occupied for centuries. McClintock found the ruts
made by Parry's cart, and was led by their appearance, after more
than forty years, to assign a very high antiquity to the Eskimo
remains. He says, "No lichens have grown upon the upturned
stones, and even their deep beds in the soil where they had rested
ere Parry's men removed them are generally distinct. The astonishing freshness of these traces compels us to assign a very considerable
antiquity to the Eskimo remains which we find scattered along the
shores of the Parry group, since they are always moss-covered, and
often indistinct."[2] I myself carefully examined several of these
traces of the wanderers, and was equally impressed with their great
age. I have here collected a list of the principal remains that have
been observed along this weary line of march :[3]—

1. The remains of huts were found by M'Clure on the north-west
coast of BANKS ISLAND.

2. On MELVILLE ISLAND Parry found the ruins of six huts, 6 feet
in diameter by 2 feet high, on the south shore of Liddon's Gulf.
Similar remains were found on Dealy Island, and at the entrance

[1] 'Blue Book,' p. 917.   [2] Ibid, p. 582. (Further papers, 1855.)
[3] Markham's 'Franklin's Footsteps,' p. 115.

of Bridport Inlet.[1]   Near Point Roche, a piece of drift timber was
seen by Vesey Hamilton, standing upright on the summit of a low,
flat-topped hill, about 300 yards from the sea, and 60 feet above its
level, but no signs of an Eskimo encampment were found near it.
The ground was covered with snow. The drift timber was 6 inches
in diameter, and was sticking up about 4 feet out of the ground,
being conspicuously placed, as if for a mark.[2]

3. BYAM MARTIN ISLAND.—Near Cape Gillman there were bones
of an ox, and jaws of a bear, and on the east shore General Sabine
saw six ruined huts and an antler.[3]

4. BATHURST ISLAND.—To the eastward of Allison Inlet there
were seven huts, some circles of moss-covered stones, and, a few
miles to the west, another hut. On the west side of Bedford Bay
there were six huts, and some circles of stones, of great age. On
Cape Capel McClintock examined ten winter habitations, and the
bones of bears and seals, some of them cut with a sharp instrument.
From various circumstances he was led to believe that none of these
huts have been inhabited within the last 200 years. The general
form of the huts is oval, with an extended opening at one end. They
are 7 feet long by 10, and are roofed over with stones and earth,
supported by bones of whales.[4]

5. CORNWALLIS ISLAND.—At the western entrance of McDougall
Bay there are some very ancient Eskimo encampments.[5] On an
islet in Becher Bay I found three moss-covered circles of stones,
the sites of summer tents, and a portion of the runner of a sledge.
West of Cape Martyr there are numerous sites of summer tents,
with heaps of bones of birds, and some very perfect stone fox-traps.
On the eastern side of Cape Martyr, Osborn carefully examined a
winter hut. Its circumference was 20 feet, and the height of the
remaining wall 5 feet 6 inches.[6] The walls were overgrown with
moss, and much skill was displayed in the arrangement of the slabs
of slaty limestone. Farther to the eastward I found traces of an
extensive winter settlement, a neat grave of limestone, and many
heaps of bones. The whole coast is strewn with remains from
Cape Martyr to Cape Hotham, and there are several on Cape Hotham
itself.

6. WELLINGTON CHANNEL.—Extensive Eskimo remains, of comparatively modern date, as compared with those at Melville Island,
were found on the extreme eastern shore, beyond Northumberland
Sound; and an Eskimo lamp was lying on the beach near Cape

¹ Parry's first voyage.   ² 'Blue Book,' p. 625. (Further papers. 1855.)
³ Parry's first voyage.   ⁴ 'Blue Book.' p. 188. (Additional papers, 1852.)
⁵ Ibid., p. 278.   ⁶ Osborn's 'Stray Leaves,' p. 143.

Lady Franklin. On the western shore of Wellington Channel, 10 miles north of Barlow Inlet, the remains of three huts were found.

7. GRIFFITH ISLAND.—I found the sites of four summer huts on the western beach, with bones of birds in and around them, also part of the runner of a sledge, a willow switch 2 feet 3 inches long, and a piece of the bone of a whale, a foot long, marked with cuts from some sharp instrument. Farther on, there were ruins of two huts, and some fox-traps.[1]

8. PRINCE OF WALES ISLAND.—On the shores of the channel, between Russell and Prince of Wales Island, there are ruins of huts, with many bones, and on the shore of a deep inlet farther west, there was an old Eskimo cache, containing bones of seals and bears.[2]

9. NORTH SOMERSET.—Ruined huts were found at Leopold Sound, and still farther south by Allen Young, who also saw semicircular walls of very ancient date, used for watching reindeer. There are now no inhabitants on North Somerset.

10. NORTH DEVON.—Remains of Eskimo huts were found on Cape Spenser, Cape Riley, and in Radstock Bay. On a peninsula at the entrance of Dundas Harbour, I found several huts with moss-covered walls three feet high, a small recess on one side, and a space for the entrance on the other. I also examined twelve tombs built of limestone slabs, containing skeletons.[3] I am aware that Eskimos belonging to the Pond's Bay tribe were afterwards met with at this place by Captain Inglefield. They had come upon the depôt which was landed at Navy Board Inlet, on the opposite coast, by Mr. Saunders, and had thence crossed over to Dundas Harbour, and finding good hunting and fishing there, they had continued to visit it in the summer. But I still think that the stone huts and tombs are the remains of a more ancient race. The Pond's Bay Eskimos, like those of Boothia and Igloolik, farther south, pass the winter in snow huts, and not in *yourts* of stone and earth.[4]

11. JONES SOUND.—An Eskimo skull was picked up.

12. CAREY ISLANDS.—Several Eskimo caches of provisions were found, in 1851, on one of the islands.

We have thus been enabled to trace the routes taken by these ancient wanderers in search of the means of sustaining life, step by

[1] 'Blue Book,' p. 266. (Additional papers, 1852.)
[2] Allen Young found remains of stones for keeping down summer huts all round the southern side of Prince of Wales Island.
[3] Markham's 'Franklin's Footsteps,' p. 61.
[4] Parry's second voyage. Ross's second voyage.

step, along the whole length of the Parry group, from Banks Island to Baffin's Bay. This region does not afford the necessary conditions for a permanent abode of human beings. Constant open water during the winter,—at all events in pools and lanes,—appears to be an absolute essential for the continued existence of man in any part of the Arctic Regions, when without bows and arrows, or other means of catching large game on land. This essential is not to be found in the frozen sea, whose icy waves are piled up in mighty heaps on the shores of the Parry Islands. Reindeer, musk oxen, and hares are in abundance on Melville and Banks Islands throughout the winter, but the emigrants, whose course we are endeavouring to trace, were no more able to catch them than are the modern " Arctic Highlanders." There animal food, too, without blubber of seal or walrus for fuel with which to melt water for drinking purposes, would be insufficient to maintain human life in the Arctic zone. As they advanced farther east they would come to the barren limestone shores of Bathurst and Cornwallis Islands, where the club moss ceases to grow, where all vegetation is still more scarce, and where animal life is not so abundant. A few years of desperate struggling for existence must have shown them that their journey half round the world was not yet ended. Again they had to wander in search of some less inhospitable shore, leaving behind them the ruined huts and fox-traps which have marked their route, and helped to identify them with the fugitives who left their *yourts* at the mouths of the Indigirka and the Kolyma. We have every reason to believe that no Eskimos have since visited the Parry Islands.

The emigrants probably kept marching steadily to the eastward along and north of Barrow Straits. They doubtless arrived in small parties throughout the fourteenth, fifteenth, and sixteenth centuries. They seem to have been without canoes, but to have been provided with dogs and sledges, and on reaching the mouth of Lancaster Sound they appear to have kept along the shore, leaving traces in the shape of ruined huts at the entrance of Jones Sound, and finally to have arrived in Greenland, on some part of the eastern shore of Smith Sound, not improbably at the " wind-loved " point of Anoritok. Thence, as new relays of emigrants arrived, they may be supposed to have separated in parties to the north and south, the former wandering whither we know not, the latter crossing Melville Bay, appearing suddenly among the Norman settlements, and eventually peopling the isles and fiords of South Greenland. Some of the wanderers remained at the " wind-loved " point, established their hunting-grounds between the Humboldt and Melville Bay glaciers, and became the ancestors of that very curious and interesting race of men, the "Arctic Highlanders."

Unlike the Parry Islands, the coast of Greenland was found to be suited for the home of the hardy Asiatic wanderers, and here at length they found a resting place. Its granite cliffs are more covered with vegetation than are the bare limestone ridges to the westward. The currents and drifting bergs keep pools and lanes of water open throughout the winter, to which walrus, seals, and bears resort. Without bows and arrows, without canoes, and without wood, the "Arctic Highlanders" could still secure abundance of food with their bone spears and darts. For generations they have been completely isolated by the Humboldt glacier to the north, and the glacier near Cape Melville to the south. Thus their range extends along 600 miles of coast-line, while inland they are hemmed in by the *Sernik-soak*, or great ice-wall. Dr. Kane tells us that they number about 140 souls,[1] powerful, well-built fellows, thick-set, and muscular, with round chubby faces,[2] and the true warm hearts of genuine hunters; ready to close with a bear twice their size, and to enter into a conflict with a fierce walrus of four hours' duration on weak ice. Their *iglu*, or winter habitation, is a circular stone hut, about 8 feet long by 7 broad, and is identical in all respects with the ruins which we found on the shores of the Parry Islands. It should be observed also that on comparing the vocabulary of the language of the Greenland Eskimo with that of the Tuski of Northern Siberia, it will be seen that both are dialects of the same mother-tongue.

The discoveries of geologists have recently brought to light the existence of a race of people who lived soon after the remote glacial epoch of Europe, and who were unacquainted with the use of metals. Their history is that of the earliest family of man of which we yet have any trace; while here, in the far north, there are tribes still living under exactly similar conditions, in a glacial country, and in a stone age. A close and careful study of this race, therefore, and more especially of any part of it which may be discovered in hitherto unexplored regions, assumes great importance, and becomes a subject of universal interest.

I ventured to hint that, after the arrival of the Asiatic emigrants at the "wind-loved" point, while some went south, and, driving out the Norsemen, peopled Greenland; and while others remained between the forks of the great glacier, a third line may have been taken far to the north, towards the Pole itself. I believe this to be

---

[1] Kane, ii. p. 108.   [2] Ibid., p. 250.

far from improbable. It is true that the "Arctic Highlanders" told
Dr. Kane that they knew of no inhabitant beyond the Humboldt
glacier, and this is the farthest point which was indicated by Kalli-
hirua—*Erasmus York* (the native lad who was on board the *Assistance*
for more than a year), on his wonderfully accurate chart. In like
manner the Eskimo of Upernivik knew nothing of natives north of
Melville Bay until the first voyage of Sir John Ross. Yet we know
that there either are or have been inhabitants north of Humboldt
glacier, for Morton (Dr. Kane's steward) found the runner of a
sledge, made of bone, lying on the beach on the northern side of
it.[1] There is a tradition, too, among the "Arctic Highlanders," that
there are herds of musk oxen far to the north on an island in an
iceless sea.[2] In 1871, during the voyage of the *Polaris*, Dr. Bessels
saw traces of Eskimos as far north as 82°, in which parallel he
picked up, lying on the beach, a couple of ribs of the walrus which
had been used as sledge-runners, and a small piece of wood that had
formed part of the back of a sledge. An old bone knife-handle
was also found, and circles of stones showing the positions of three
tents of a summer encampment. Assuredly the greater abundance
of game far up Smith Sound, as described by Dr. Bessels, shows
that the Eskimos who wandered towards the Pole would have no
inducement to go south again. Open water means to them life.
It means bears, seals, walrus, ducks, and rotches. It means health,
comfort, and abundance.

In the belief of some geographers there is a great *Polynia*, or
basin of open water round the Pole.[3] Wrangell says that open
water is met with north of New Siberia and Kotelnoi, and thence
to the same distance off the coast between Cape Chelag-koi and
Cape North.[4] If this be the case the Omoki and Onkilon, who fled
before Tartar or Russian invasion, had no reason to regret their
change of residence. A land washed by the waves of a Polar Sea
would be a good exchange for the dreary *tundra* of Arctic Siberia,
where the earth is frozen for 70 feet below the surface. Wherever
a *Polynia*, be it large or small, really exists, there men who sustain
life by hunting seals and walrus may be expected to be found upon
its shores. We may reasonably conclude then, if the region
between Hall's farthest and the Pole bears any resemblance to
the coast of Greenland, if there is a continent or a chain of islands
with patches of open water near the shores, caused by ocean currents, that tribes will be found resembling the "Arctic High-

---

[1] Kane, i. p. 309.   [2] Petermann's 'Search for Franklin.'
[3] Hayes, p. 35.   [4] Wrangell, p. 504.

landers," who extend their wanderings to the very Pole itself. Such
a people will be completely isolated, they will be living entirely
on their own resources—far more so even than the "Arctic Highlanders," since the North Water has been for the last forty years
visited by whalers and explorers: and a full account of the habits,
the mode of life, and the language of so isolated a people will be
to many of us among the most valuable results of the contemplated
Arctic Expedition.

I have thus endeavoured to point out the routes which were
probably taken by the ancestors of the Greenlanders, and of the
supposed denizens of the Pole, in their long march from the Siberian
coast.

2.

ON THE ARCTIC HIGHLANDERS.

THE country of the Arctic Highlanders, the most northern known
people in the world, is that strip of land on the eastern side of Baffin's
Bay and Smith Sound, which is bounded on the south by the
Melville and on the north by the great Humboldt glacier; and in
describing a strange and very interesting tribe, it will be well, in the
first place, to enumerate the voyages which have brought this region
to our knowledge, and to examine what manner of country it is
which supplies a home for this outlying piquet of humanity.

On the 1st of July, 1616, Baffin steered the little *Discovery*, of
fifty-five tons, into the open water at the head of Baffin's Bay,
which "anew revived the hope of a passage." The old navigator
refrained from scattering the names of all the great men of his
day and of all his friends and acquaintances round the head of
the bay. He only gave names to nine of the most prominent
features—namely, Cape Dudley Digges, Wolstenholme Sound and
Island, Whale Sound, Hakluyt Island, the Carey Islands, and Smith,
Jones, and Lancaster Sounds. He anchored in Wolstenholme and
Whale Sounds; but it is not stated that he landed, and as the weather
was bad, he probably did not, but he communicated with the inhabitants. No doubt, too, they were watching him with extreme
astonishment, from behind rocks, as is their wont, and the appearance of this strange apparition in those silent seas may have
been the subject of a tradition in the tribe.

Baffin, then, was the first navigator who forced his way through the
ice-barrier drifting south, and entered the "North Water;" but it
was left to Sir John Ross to discover the existence of inhabitants on

its shores. His account of them, though containing several errors, is given in perfect good faith, and due allowance must of course be made for mistakes of interpretation.

After an interval of just two centuries, Captain John Ross followed Baffin into the "North Water," and was the first European who had intercourse with the inhabitants of its shores—whom he called "Arctic Highlanders." They came off to his ships over the ice, in small parties, between the 9th and 16th of August, 1818, and he took much pains to obtain all possible information from them, through his Eskimo interpreter, John Sackheuse; but he did not land to examine their huts. Sackheuse evidently understood their dialect very imperfectly, and he told Ross strange stories about a mountain of iron, a king called *Tuloowah*, who lived in a large stone house, and other marvels. But all that Sir John saw with his own eyes, respecting the dress and appearance of his visitors, their sledges and implements, he describes with truth and accuracy.

Sir John Ross led the way into the "North Water," and he was followed during many years by a fleet of whalers who, doubtless, occasionally communicated with the "Arctic Highlanders;" but we have no record of these visits, if any such took place. In 1849-50 the *North Star* (store-ship) wintered in Wolstenholme Sound, and her crew had most friendly relations with the natives throughout the period of their stay; and in August, 1850, H.M.S. *Assistance* (Captain Ommanney), with her tender, the *Intrepid*, communicated with the natives at Cape York. The *Intrepid* also went into Wolstenholme Sound; and we took on board a young Arctic Highlander, of whom I shall have more to say presently, as he afforded an excellent opportunity of forming a judgment of the characteristics of this interesting people. The other discovery ships of 1850-51 (*Lady Franklin* and *Sophia*, under Captain Penny: *Prince Albert*, under Captain Forsyth; and *Felix*, commanded by Sir John Ross) also had intercourse with the natives at Cape York. In August, 1852, H.M.S. *Resolute* (Captain Kellett) touched at Cape York; and in the same year Captain Inglefield, in the *Isabella*, visited the natives of the Petowak glacier, and at a settlement about twenty miles from Cape Parry. Dr. Kane did not see them until his schooner was frozen in for the winter on the eastern shore of Smith Sound, but he afterwards formed most intimate relations with them during 1853-54-55. One of his officers, Dr. Hayes, was living amongst them for several months, and they saved the lives of Kane and his whole crew. Sir Leopold McClintock, in the *Fox*, communicated with eight natives off Cape

York, on June 27th, 1858. They asked after Dr. Kane, and immediately recognised the Danish interpreter, Petersen, who served both in the expeditions of Kane and McClintock. At Godhavn Sir Leopold received a request from the Royal Danish Greenland Company, through the Inspector of North Greenland, to convey the tribe of "Arctic Highlanders" to the Danish settlements in Greenland; and, he says, "had the objects and circumstances of my voyage permitted me to turn aside for this purpose, it would have afforded me very sincere satisfaction to carry out so humane a project."[1] Dr. Hayes saw much of them again during his voyage in 1860, as did Dr. Bessels and the crew of the *Polaris*, when they wintered off Etah in 1872–73.

It is from the accounts of writers and other observers who have served in these different voyages, and more particularly from the works of Dr. Kane and Dr. Hayes, that our knowledge of the " Arctic Highlanders " is derived.[2]

The home of these people of the far north is between latitudes 76° and 79°, just on the verge of the unknown Polar Region. It is a deeply indented coast-line of granitic cliffs, broken by bays and sounds, with numerous rock sand islands, and glaciers streaming down the ravines into the sea. To the south it is bounded by the glaciers of Melville Bay, which now bar all progress in that direction, insomuch that when John Sackhouse told Captain Ross's visitors that he came from the south, they replied—"that cannot be, there is nothing but ice there."[3] To the northward, in like manner, a glacier bounds their hunting-ground; while inland the mighty *Sernik-soak*, or great glacier of the interior, confines them to the sea-coast, and to the shores of fiords and islands. The vast interior glacier sends down numerous branches to the sea, the ends of which break of and form a great annual harvest of icebergs. The rocky coast, between these streams of ice, is for the most part of granite formation, and in many places is richly

---

[1] Fate of Franklin, p. 138.

[2] 1. Sir John Ross's First Voyage, 1818; 2. Parker Snow's Arctic Voyage, 1851; 3. Osborn's Stray Leaves, 1852; 4. Markham's Franklin's Footsteps, 1853; 5. Sutherland's Journal, etc., 1852; 6. Inglefield's Summer Search, 1853; 7. Arctic Miscellanies, 1853; 8. McDougall's Voyage of the *Resolute*, 1855; 9. Kane's Arctic Explorations, 1856; 10. Hayes' Boat Voyage, 1857; 11. 'Rev. J. B. Murray's Account of Erasmus York; 12. McClintock's Fate of Franklin, 1860; 13. Hayes' Narrative of a Voyage towards the North Pole in the schooner *United States*, 1867. *Vocabularies*—1. Balbi, Atlas Ethnographique; 2. Washington, Eskimo Vocabularies; 3. Fabricius, Greenland Dictionary; 4. Ross's Second Voyage; 5. Parry's Second Voyage; 6. Crantz's Greenland; 7. Egede's Greenland and Janssen's Vocabularies. *Siberia*—Strahlenburg; Wrangell; Hooper's Tents of the Tuski; Dr. Simpson's Report.

[3] The distance from Cape York to Upernivik, the nearest inhabited land to the south, is about two hundred and fifty miles.

covered with soft moss, and numerous wild flowers, besides dwarf willow. The flora of this land consists of forty-four genera and seventy-six species as yet discovered, among which there are four kinds of ranunculus, fourteen crucifers, including three kinds of scurvy grass, several pretty little stellarias, potentillas, and saxifrages, seven of the heath tribe, a dwarf willow, a fern (*Cystopteris*), and numerous mosses and grasses. Dr. Donnet speaks of the fertile valleys of Wolstenholme Sound, covered with moss, over which, as he walked, he felt as if Persia had sent her softest material to give comfort to the "Arctic Highlander." It is fair to add that he wrote this sentence when frozen in off the more barren shores of Griffith Island.

But it is on the condition of the sea, much more than of the land, that the suitability of a region for human habitation depends within the Arctic Zone; and although Greenland is infinitely richer in vegetation, and abounds more in animal life, than the dreary archipelago to the westward, yet without open water in the winter it would be uninhabitable. The ice drifting south in the spring leaves a large extent of navigable sea at the head of Baffin's Bay during the summer—known as the "North Water"; while the currents and the innumerable icebergs, always in motion and ploughing up the floes, keep up open pools and lanes of water throughout the winter.

Such is the country which supports a multitude of living creatures, in a temperature where the mean of the warmest month is + 38, and of the coldest − 38, in a climate where there are furious gales of wind, where the year is divided into one long day and one long night, but where, in the glorious summer, in the calm and silent sunny nights, may be seen some of the most lovely scenery on this earth. No rich woodland tints, little diversity of colouring; all its beauty dependent upon ice and water, and beetling crags, and strange atmospheric effects, but still most beautiful. The land between the shore and the glacier is the abode of reindeer, bears, foxes, and hares; of ravens, falcons, owls, ptarmigan, willow-grouse, snow-bunting, dotterels, and phalaropes; while the aquatic birds come in tens of thousands to breed on the crags and islands—king ducks, eider ducks, longtailed ducks, and brent geese; looms, dovekeys, and rotches in millions; skuas, ivory and silver gulls; burgomasters, mullemukkes, kittiwakes, and Arctic terns. Above all, so far as man's existence is concerned, the open pools and lanes of water are crowded with seals (hispid and bearded), walrus, white whales, and narwhals, and these again betoken the existence of fish, molluscs, and minute marine creatures in myriads.

Here, then, is a region where man too might find subsistence, and here accordingly we meet with a hardy tribe of men, numbering, according to Dr. Kane's calculation, about 140 souls, reduced, according to Hayes, to 100 in 1860. They are separated in eight or more settlements, scattered along the coast from the Humboldt to the Melville glacier. The names of the settlements, according to York, who marked all their positions on his chart, are *Anoritok*, in Smith Sound; *Etah*, near Cape Alexander; *Pikierlu*, *Ekaluk, Pitorak, Natsilik*, in Whale Sound; *Umenak*, where the *North Star* wintered; *Ahipa* and *Innagen*, at Cape York. These are the permanent winter settlements, but in summer they pitch their tents wherever they are likely to find the best hunting-ground.

This remarkable tribe is decidedly of Asiatic affinities so far as the outer man is concerned. The men we saw at Cape York averaged about five feet five inches in height; but Dr. Kane describes the first native he met with as a head taller than himself, and extremely powerful and well built. They are generally corpulent and fleshy, and so heavy that it is difficult to lift a full-grown man. The forehead is narrow and low; nose very small; cheeks full and chubby; mouth large, lips thick; eyes small, black and very bright; beard scanty, and hair black and coarse. The hands and feet are small and thick. They are possessed of great strength, endurance, and activity; and are on the whole intelligent. This description, most of which I have copied from my journal, would answer as well for some of the northern tribes of Siberia as for the Arctic Highlanders; and I may add that when poor York went to the Great Exhibition, everybody thought he was a Chinese.[1]

Their winter habitations mark them as a peculiar people, in some respects distinct from the Eskimo of America; for while the latter live in snow huts, the Arctic Highlanders build structures of stone. These stone *iglus*, though quite unlike the winter homes of the American Eskimo, are precisely the same as the ruined *yourts* on the northern shores of Siberia, and as the ruins found in all parts of the Parry Islands. They thus furnish one of several clues which point to Siberia as the original home of these people.

The *iglu* of the Arctic Highlander is built of large stones, carefully and artistically arranged in an elliptical form. The sides gradually approach each other, and the roof is covered over

---

[1] The descriptions given by Dr. Simpson of the tribes in Kotzebue Sound, and by Lieut. Hooper of the Tuski on the Asiatic coast, show that these people closely resemble the Arctic Highlanders in outward appearance.

with long slabs, at a height of about five feet eight inches from the ground, the outside being lined with sods. The entrance is by a tunnel about ten feet long, with barely room enough for a man to crawl through—called *tossut ;* and just above there is a small window with dried seals' entrails stretched over it. The dimensions of the interior are about twelve feet by ten, and half of it is taken up by a raised platform which is covered with dried moss and bear-skins, and serves as a bed for the whole family. On the walls hang skins, fowl-nets, whips, and harpoon-lines ; and the furniture consists of shallow cups of seal-skin, the soap-stone lamp (*kotluk*) with its supply of oil and moss-wicks, and racks of rib-bones lashed together crosswise, on which the clothes are dried. The cups are for receiving the water as it melts from a lump of snow, and flows down the shoulder-blade of a walrus, placed on stones. This is their sole cooking operation; for the boiling of soup made of blood, oil, and intestines is only done as an occasional delicacy ; and as a rule they devour their food raw, be it flesh, blubber, or intestines, and in enormous quantities. Kane calculates one man's consumption at eight or ten pounds of flesh and blubber, and half a gallon of water and soup. This diet is no doubt wholesome and natural, and, so long as it can be had in sufficient quantity, it preserves the Arctic Highlander in the fine plump condition which characterises him. The heat of the *iglu* is intense when the ordinary number of a dozen inmates is collected, and it is the usual habit to adopt a complete dress of nature as the indoor attire. It is not, therefore, until the Arctic Highlanders come forth for the chase that they may be seen in a dress suited to the outer climate. Next the skin they wear a shirt of bird-skins neatly sewn together, with the soft down inwards ; over which comes the *kapetah,* a loose jumper of fox-skin, which is, however, tight round the neck, where the *nessak* or hood is attached to it. The *nessak* is lined with bird-skins and trimmed with fox-fur. The breeches, called *nannuk,* of bear-skin come down to the knees, and up so as just to be in contact with the *kapetah* when the wearer is standing upright. If he stoops the whole of his person between the *nannuk* and *kapetah* is exposed. On the feet bird-skin socks are worn with a padding of grass, over which come bear-skin boots. By means of their sledges drawn by dogs they can move swiftly to the best hunting-grounds, which are of course well known, and secure the mighty game, the huge walrus and formidable bears, which are their necessaries of life. No hunters in the world display more indomitable courage and presence of mind, nor more skill and judgment in the exercise

of their craft. Their weapons are a lance of narwhal ivory, or
sometimes of two bear thigh-bones lashed together, tipped with
steel since their intercourse with whalers, and a harpoon. They
also have a knife made from some old drifted cask hoop, which
they conceal in the boot. The lance is used in their gallant encounters with bears, and in securing a walrus or seal on the ice,
when its retreat has been cut off; the harpoon for the far more
dangerous battles with the walrus in his own element. They
have bird-nets, with which they catch the little auks and guillemots that breed in myriads on the perpendicular crags; and
this employment is also attended with great risk. In the year
we visited Cape York, a native told us that several men had lost
their lives in netting guillemots on the steep cliffs of Akpa Island.
York also told us that his people occasionally, but very rarely,
succeeded in killing a reindeer; and Petersen says that the twenty
decayed skulls, without lower jaws, that were found in the northward of 79° N., had been killed by native hunters.[1]

They have no canoes, either *kayak* or *umiak*, and are thus confined to the land and ice; and they probably first obtained the
word *umiak* for a ship, from John Sackhouse when he pointed
to the *Alexander* and *Isabella*. This ignorance of an appliance
which is known to nearly all the Eskimo tribes is remarkable.
The Arctic Highlanders certainly do not show themselves to be
less intelligent than other Eskimo tribes in contrivances for procuring food and providing for their comfort. I am inclined, therefore, to account for their want of *kayaks* from the circumstances of
their position. In the south, from the absence of ice during a great
part of the year, the Greenlander is obliged to seek his food on
the sea; while in the north there is a land-floe throughout the
year, and the Arctic Highlander can harpoon the walrus, narwhal,
and white whale from the ice. The necessity which led to the
invention of a *kayak* in the one case, does not exist, in so urgent
a form, in the other. Hans, the Holsteinborg Eskimo, who was
left behind by Dr. Kane (having fallen in love with a fair
daughter of the far north), had a *kayak* with him; but in the
winter of 1857–58, being pressed by famine, he and his family
were obliged to eat it.

It is more remarkable that the Arctic Highlanders have no
bows and arrows, and this is one of the circumstances which con-
clusively prove that they are not the same people as the Eskimo
of Boothia and Pond's Bay. The great superiority of the sledges

_____

[1] McClintock, p. 76

of the Arctic Highlanders, compared with those of the Boothia people,[1] must weigh on their credit side, against the bone bows and arrows, in deciding the comparative ingenuity and intelligence of these tribes.

The hunting season of summer and autumn enables the Arctic Highlanders to accumulate large stores of flesh and blubber which last them until December, but their enormous consumption soon diminishes the stock, and in January and February they begin to feel the pinchings of hunger. Then these indomitable hunters have to come out in the intense cold and contend with the huge walrus on the edge of treacherous ice; while, in very bad seasons, they are reduced to eating their dogs. During the long night they are engaged in mending sledge-harness and preparing harpoon-lines and bird-nets; and the women chew the boot-soles and bird-skins, and make clothes with ivory needles and thread of split seal sinew. Summer brings a bright and happy time of sunshine and plenty. The children drive the babies along in miniature sledges, the boys play at hockey with rib-bones and leathern balls, or catch the rotches with nets attached to long narwhal horns, and the hunters are busy in their attacks upon larger game. All emerge from the dismal *iglus*, and exchange their darkness and filth for the well-ventilated seal-skin tents; and thus they move from place to place along the coast.

We now come to the consideration of the important question of language as an element in the discussion of the origin of these people. That of the Greenland Eskimo belongs to the American type of languages, which Du Ponceau has called polysynthetic, and William von Humboldt agglutinative, from their peculiarity of forming all compound words and phrases by adding particles to the root in a certain way. The Eskimo language certainly does exhibit this peculiarity. It indulges in very long words, such, for instance, as *Aulisariartorasuarpok* (he made haste to go out fishing), which is composed of the three words, *aulisarpok* (he fishes). *peartorpok* (he went), and *picesuarpok* (he made haste). *Aglekkigiarto-rasuamiarpok* is not short. But agglutination is by no means peculiar to the American languages; and Professor Max Müller groups the American with many other languages in Asia and Africa, which he calls agglutinative, not because there is the remotest indication of a common origin, but from the absence of any organic differences of grammatical structure.[2] This is, there-

---

[1] McClintock., p. 236.

[2] These languages are called agglutinative, to distinguish them from the inflexion of the Aryan and Semitic tongues, and from the roots of the Chinese.

fore, not a conclusive reason for supposing that the Eskimo is an American language. The vocabulary of the Greenland language and that of the Tuski tribes of Siberia contain so many important words alike that their comparison supplies a strong argument in favour of common origin. The following list contains some words which are identical in the languages of the Greenlanders and of the Siberian tribes near the Gulf of Anadyr :—

ENGLISH.	GREENLANDERS.	SIBERIAN.
Sun	Sekkinék	Shekenek.
Earth	Nuna	Nuna.
Water	Imak	Imak.
Fire	Ing-nek	Eknok.
Father	Atatak	Ataka.
Eye	Ieé	Iik.
Head	Niakok	Naskok.
1	Atausek	Atashek.
2	Marluk	Makukh.
3	Pingasut	Pingayu.
4	Sisamet	Ishlamat.
5	Tellimet	Tatlimat.

But neither the Arctic Highlanders' vocabulary nor that of the Tuskis and Anadyr tribes are before us in a complete shape.[1] It may be understood generally that the languages spoken by all the tribes from Humboldt Glacier to Cape Farewell, are but dialects of the same mother-tongue ; while they are dialects of the languages of Labrador, Igloolik, Boothia, Kotzebue Sound, and some parts of Siberia.

The Arctic Highlanders only have words for the first five numerals, although they make shift to count a little higher, up to twenty ; but otherwise their language, though wanting all words to express abstract ideas, is very precise and exact, and few languages are richer in pronominal forms of speech. Their songs are for the most part impromptu, and in the long winter night, while one recites a catalogue of recent events and possibly some traditions, the rest join, with a certain time and cadence, in the ancient chorus — *Amna ajah ajah ah-lu*. Dr. Kane heard this chorus in the *iglus* in Smith Sound, and Crantz records the same words as used by the people of South Greenland. Their religion is very simple. They believe in supernatural beings presiding over the elements, who are the familiar spirits of their *angekoks* or magicians ; and that the *angekoks* can converse with them, and thus prophecy the prospects of the hunting season and similar matters.

---

[1] It must be remembered, too, that the Omoki and other Siberian tribes have disappeared altogether, taking their language with them ; and, according to my theory, these are the ancestors of the Arctic Highlanders.

These *angekoks* are not hereditary office-bearers ; but, for the most part, they are the cleverest and laziest fellows in the community. They have a few proverbs and figurative sayings, they perform incantations over the sick, prescribe the nature and amount of mourning for the dead (who are buried under heaps of stones, or sometimes an *iglu* is abandoned and closed up as a tomb), and exercise that general influence which they obtain from their own cunning, and from the traditional respect in which their profession is held. This *angekok* superstition is exactly the same as the Shamanism of the Siberian tribes, as described by Wrangell. There is very little crime amongst these good-natured savages, though the punishments they inflicted on criminals were formerly severe. But, in 1858, the people at Cape York told McClintock ' that they had abolished their ancient custom of punishing theft capitally, because their best hunters were often the greatest thieves.

One of the most striking points in the intellectual development of all the Eskimo tribes is their wonderful talent for topography. The cases of the woman of Igloolik who drew a map for Parry, and of the Boothians who did the same for Ross, and the interesting account of the old lady who "conned" the *Fox* up Pond's Inlet as if she had been a certified pilot from the Trinity House, are familiar to the readers of Arctic voyages. The same talent was displayed by our shipmate, Erasmus York, on board the *Assistance*. When asked by Captain Ommanney to sketch the coast, he took up a pencil, a thing he had never seen before, and delineated the coast-line from *Pikierlu* to Cape York, with astonishing accuracy, making marks to indicate all the islands, remarkable cliffs, glaciers, and hills, and giving all their native names. "Every rock," says Dr. Kane, "has its name, every hill its significance."

The visitor who first sees a party of Arctic Highlanders will be at once struck by their merry, good-natured countenances, their noisy fun, and boisterous laughter. They have a true love of independence and liberty, and their mode of life has bred in them great powers of endurance, cool presence of mind, and indomitable courage. Their ingenuity and skill are by no means contemptible, and their intellectual capacity, though inferior to that of many other savage people, is not altogether despicable. They do not hesitate to steal from the stranger, for whom they cannot be expected to have any fellow-feeling; but when confidence is once established, they have proved themselves to be good men and true ; they undoubtedly saved the crew of the *Advance* from death, which was staring them in the face ; and Dr. Kane gives his testimony that

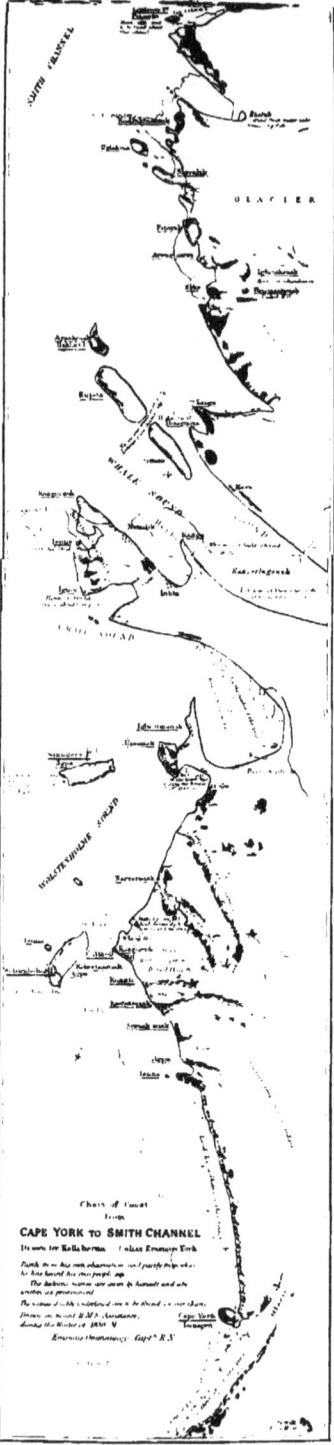

Chart of Coast
from
CAPE YORK to SMITH CHANNEL

" when troubles came upon him and his people, never have friends
been more true than these Arctic Highlanders."

We, of the old *Assistance*, can bear witness with regard to one of
the members of this northern race, who, by his constant cheerful-
ness and good humour, and his readiness to make himself useful,
became a great favourite on board. Through the kindness of
Admiral Ommanney he received an education in England, and
went afterwards to Newfoundland, where he died in 1856. A lady,
who wrote to announce his death, thus speaks of poor Erasmus
York (Kallihirua): " During his illness he was as patient and
gentle as ever, and thankful for all that was done to relieve him.
We all loved him for his true-heartedness, obedience, and kindness of disposition ; and I trust that we may not forget the example
he gave us of forgiveness and forbearance under injury." The
Arctic Highlanders are savages, but they are ingenious and intelligent—courageous as hunters, true and loyal to friends in distress,
and capable, after instruction, of the highest virtues of civilised men.

In conclusion, I will sum up the points which, after an examination of the ethnology of the Arctic Highlanders, tend to corro-
borate my theory of their origin and migrations. First, then, there
is the evidence that they are not branches of any Eskimo tribe of
America or its islands. The American Eskimos never go from
their own hunting range for any distance to the inhospitable north.
Except in the case of the Pond's Bay natives, who followed up the
whalers for a specific reason in modern times, there is no instance
of their having gone north ; and it is unreasonable to suppose that
they would do so. The American Eskimos live in snow huts, the
Arctic Highlanders in *iglus* built of stone ; the former have *kayaks*
and bows and arrows, the latter have none; the Boothians use
sledges of rolled-up seal-skin ; the Arctic Highlanders have sledges
of bone. We have proofs, also, that the ancient wanderers who
left traces along the Parry Islands, were the same tribe as the
Arctic Highlanders, and distinct from the American Eskimos.
The ruins on the shores of the Parry Islands are identical with the
stone iglus of the Arctic Highlanders, and unlike the habitations
of the American Eskimos. The pieces of bone sledge-runners
that were found among these ruins, are the same as those used by
the former tribe, while the Boothians (the nearest American
Eskimos) use seal-skin runners. The bone which had been cut to
form a duct for conducting melted snow into a cup, found by
myself on Griffith Island, and the lamp picked up by Osborn on
Cape Lady Franklin, are precisely similar articles to those now
used by the Arctic Highlanders.

We now come to the points of resemblance between the Arctic Highlanders and some Siberian tribes. In physiognomy and general appearance the Eskimos are unlike any other American people. The Eskimo language in vocabulary and grammatical construction is but a dialect of the language spoken by the Tuski, the people in the Gulf of Anadyr in Siberia. The *angekok* superstition of the Eskimo resembles, even in minute particulars, the Shamanism of Siberia. These points apply to the whole Eskimo race; and, in proving that the Arctic Highlanders are distinct from the American Eskimo, I do not mean that they are not all the same race, speaking dialects of the same language, but that they have had no communication since their ancestors left Siberia, and, crossing the meridian of Behring Straits, wandered to the northward and eastward. The American Eskimo migrated at some very remote period, from Siberia by way of Behring Strait; and the Viking Thorwald found them on the coast of Labrador in the tenth century. The migrations from the northern coast of Siberia were later, and were caused by Central Asiatic encroachments from the eleventh to the fourteenth centuries. This exodus took a distinct and more northern route, along the coast of the Parry Islands, to Greenland. Such are the proofs that have convinced me that the cradle of the Eskimo race is to be found on the frozen tundra of Siberia.

Only a small remnant of these ancient wanderers is represented by the Arctic Highlanders; and, as I have already suggested, many parties as they arrived, continued their journey to the south, where they peopled Greenland. Others probably took a more northern course. As to the Greenland population, the historical testimony of the Norsemen, and the universal tradition of the Greenlanders themselves, unite in affirming that the first Skrœllings came from the north.[1] This is not the place for critically discussing the value of ancient Icelandic records, which have been most ably edited by learned Danes. From them we learn that the Norsemen were the first inhabitants of Greenland, and that the present population first appeared, coming from the north, in the fourteenth century. How it was that the diminutive, though muscular and courageous, Skrœllings overcame and annihilated their gigantic Scandinavian foes, must for ever remain a mystery. It may be that the Normans were first thinned down by disease, and greatly reduced in numbers. One thing is certain: the Normans disappeared, leaving

---

[1] John Sackhouse, when he first saw the Arctic Highlanders, immediately mentioned the universal tradition of his people that they originally came from the north.

many ruins and runic inscriptions behind them, and the Skrællings have taken their place. The modern Danish Eskimos have detailed traditions of the wars between their ancestors and the *Kablunu*, which are represented in the curious woodcuts brought home by Sir Leopold McClintock, and have since been published by Dr. Rink, who, I understand, is of opinion that these Eskimo traditions are founded on historical facts.

But it is the northern, and not the southern, migration of the Arctic Highlanders that now demands our attentive consideration. We here approach the very confines of the great unknown polar region, and we can discover indications of the existence of a polar population up to the very threshold of the *Terra Incognita*. Petersen tells us that he saw ruins of stone *iglus* to the northward of latitude 79° N., which were evidently upwards of two centuries old ; and the runner of a sledge was picked up beyond the Humboldt Glacier. Dr. Bessels found Eskimo remains still further north. Here, then, are the traces of wanderers coming south from the Polar region. Clavering, in 1823, met with two families in the most northern part of East Greenland, who must have come from the north, and have wandered completely round the still unknown northern shores of the great glacier-bearing continent of Greenland.[1]

These people had wandered away, or died out, when the German Expedition visited the same part of the coast in 1869–70, but numerous remains of their sojourn were found, consisting of graves, and *iglus* or huts.[2] Much farther south, on the east coast of Greenland, Captain Graah, in 1829, found several places inhabited; and he gives a very interesting account[3] of these East-landers, as he called them, who in 1830 numbered not more than 480 souls, in twelve different localities along the coast, from the Danebrog Islands to Cape Farewell. The Eastlanders also came from the north, and not from the west side round Cape Farewell.

There is thus sufficient proof that people have reached the east coast of Greenland from the north, and, consequently, that they have wandered for many hundreds of miles over the unknown area. It is certain that their remains will be found, and if there are *polynias* of open water, as at the north end of Baffin Bay, it is probable that there are still inhabitants at the North Pole or near it.

[1] We may infer that they did not come from the south, for the same reason that the American Eskimo have never gone north to the Parry Islands. The East Greenland coast, from the Danebrog Islands to Hudson's " Hold with Hope," is so blocked up with eternal ice that no human being could exist there, certainly none would wander there, from the more genial south.

[2] See ' German Arctic Expedition ' of 1869–70, chapter xiv.

[3] See ' Narrative of an Expedition to the East Coast of Greenland,' by Capt. W. A. Graah (Murray, 1837), pp. 111–121.

They will have taken a route from Siberia to the north of the Parry Islands; while another division of the wanderers passed along the southern shores, to the region between the Melville and Humboldt glaciers. But man is not the only animal that has journeyed round the northern side of Greenland. The musk-ox is not known in the inhabited parts of that region. It is a peculiarly American form. Yet the crew of the *Polaris* found it up Smith Sound, and the Germans met with it on the east coast. The little *Mus Hudsonicus* is also American, and unknown in West Greenland, and it also was found by Dr. Bessels. These are direct and positive proofs of migrations along the northern face of Greenland from the American side, and of an inhabitable region, capable of supporting very large ruminants, within the unknown area.

The problems thus indicated are among the most interesting that will occupy the attention of the Arctic Expedition. In the possible, if not probable, event of a new people being discovered, a list of words in ordinary use among their distant kindred in West Greenland will be needed for comparison. A sketch of the grammar from Crantz and Janssen, and some vocabularies, have therefore been prepared. The vocabularies have been collected from Admiral Washington's little book,[1] with additions from Crantz, Kane, Janssen,[2] and Kleinschmidt.[3]

NAMES OF ARCTIC HIGHLANDERS.

(From Kane, Hayes, and Bessels.)

Akomodah (K.), a fat boy in 1854, son of *Metek.*
Alatah (H.).
Amalutok (K.), half-brother of *Metek.*
Anak (K.), wife of *Nessak.*
Angeit (H.), "the catcher." Son of *Kablunet.* Brother of Mrs. Hans.
Aningnak (K.), wife of Marsumah.
Arko (H.), "spear thrower." A boy of 12 in 1860.
Aronnelik (K.), wife of Tellerk.
Awahtok (K.).
Cheichenguak (H.).
Irki (K.).
Itukichu (B.), a good hunter.
Ivalla (B.), wife of Itukichu.

---

[1] Admiral Washington's vocabulary of Greenland Eskimo was drawn up for him by Mr. Nösted, a Danish Missionary, in 1852; and every word was gone over and revised by Erasmus York (Kalli-hirua), under the supervision of the Rev. Henry Bailey, Warden of St. Augustine's College at Canterbury, and of Dr. Rost, then Professor of Sanscrit at that college.

[2] 'Elementarbog i Eskimoernes Sprog til brug for Europæerne ved Colonierne i Grønland, bed,' E. C. Janssen. (Kjobenhavn, 1862.)

[3] 'Grammatik der Grönländischen Sprache mit theilweisem Einschluss des Labradordialects,' von S. Kleinschmidt. (Berlin, 1851.)

Kablunet (H.), " white skin." Wife of *Tcheitchenguak.* Died in 1860. Mother of Mrs. Hans.
Kalutah (K.).
Kalli-hirua, or " Erasmus York." Came on board H.M.S. *Assistance,* 1850, at Cape York. At St. Augustine's, Canterbury. Died at St. John's, New-foundland, in 1856.
Kalutunah (K. and H.), the Angekok, and, in 1860, Nalegak of the tribe. The best hunter.
Kartak H.), a girl engaged to *Arko* in 1860.
Kesarsoak (H.), " white hairs." The oldest hunter in the tribe in 1860.
Kresut (K.), " driftwood." A blind old man.
Marsumah (K.).
Merkut (K.), wife of Hans. Daughter of Shang-hu.
Metek (K.), " eider duck." Chief of Etah in 1854.
Myuk (K. and H.), son of *Metek.* A loafer. One of Satan's light infantry.
Nessak (K.), " jumper hood."
Nualik, né *Eguok* (K.), wife of *Metek.*
Paulik (K.), nephew of *Metek.*
Piugasuk (H.), " the pretty one." Child of Hans.
Shang-hu (K.).
Sipsu (K. and H.), " the handsome boy," murdered by *Kalutunah.*
Tatterat (K. and H.), " Kittiwake." Always out at elbows. A loafer.
Tellerk (K.), " right arm."
Utaniak (K. and H.)

NOTE ON THE ORTHOGRAPHY.

A is to be sounded as in *father* when long, as the *u* in *but* when short. *E* as in *there,* *i* as in *ravine,* *o* as in *more,* *u* as in *flute,* *ai* as *i* in *time,* *au* as *ow* in *how,* *ok* as *oak.* *Ch* as in *church,* *g* as in *get,* *kh* as *ch* in *loch.*

The word *Esquimaux* is a term of the Dog-rib Indians, meaning " flesh eaters," and was first given to the Innuit by the French Canadians, whence the strange orthography. The simpler and proper form, adopted by the Danes and by Admiral Washington, is *Eskimo.*

3.

LANGUAGE OF THE ESKIMO OF GREENLAND.

SKETCH OF THE GRAMMAR.

The first Eskimo grammar was by Hans Egede, published in 1760. The second, written by Königseer, in 1780, is still in manuscript. That of Fabricius, who long resided in Greenland as a missionary, appeared in 1791. Kleinschmidt published his Eskimo grammar at Berlin, in 1850; and the vocabularies of Janssen appeared at Copenhagen in 1862.

The Eskimo language belongs to the American group; the nouns are declined by the addition of terminations to the roots, and the adjective follows the substantive. There is a great abundance of modes of expression effected by processes of agglutination, the particles conveying various meanings, and modifications of meanings.

There is no article, and the dual and plural nouns are formed by
the addition of particles; thus:—

Singular.	Dual.	Plural.
Nuna	Nuneik	Nuneit, *land.*
Uyarak	Uyarkak	Uyarket, *stone.*
Iglu	Igluk	Iglut, *house.*
Innuk	Innuk	Innuit, *man.*

Collective nouns have only the plural, and end in "*it*," as *Iglu-
perksuit*, a collection of houses.

The genitive is formed by the addition of *b*, or of *m* if a vowel
follows. The other cases are formed by adding the following particles, acting as prepositions:—

Mik, with, at, through.
Mit, from.
Mut, to.
Mi, in, on.
Kut, through, over.
Agut, round.
At, underneath.
Kut, above.
Suk, beyond.
Tuno, behind.

Kingo, further end.
Ake, opposite.
Iluk, inner.
Silat, outer.
Avut, farthest.
Kit, seaside.
Kange, landside.
Kujat, south side.
Avanguek, north side.

(Where?)	Nunama,	on land.
(Whence?)	Nunamit,	from land.
(Whence?)	Nunakut,	over land.
(Where?)	Nunamut,	to land.
(How?)	Nunamik,	with land.
(Time)	Ukinar,	in winter.
(Time)	Unukut,	in evening.
(Mode)	Okautsinik,	with words.

PARTICLES.

The meanings of nouns are also varied in numerous ways by the
addition of particles. Of these the most common are the augmenta-
tives and diminutives. *Suak* or *suit* means great; as *Nuna*, land;
Nunarsuak, great land. *Kingmek*, a dog; *Kingmersuak*, a great dog.
Ngnak, is small; as, *Kingminguak*, a small dog. *Gusak* denotes what
belongs to or is part of anything; as, *Umiak*, a boat; *Umiagasak*,
what belongs to a boat. *Inak* and *tuak* are only. as *Iglu*, a
house; *Igluinak*, one house only. *Ernek*, a son; *Ernituak*, an only
son. *Siat* denotes ordinary size, neither large nor small; as, *Kakak-
siat*, a middle-sized bill. *Liak* is a particle which denotes that the
thing indicated by the noun to which it is attached was made by
its owner as *Iglerfik*, a box; *Igerfiliak*, his box made by himself. *Siak* implies that the thing was bought as *Savik*, a knife;
Saviksiak, a purchased knife. *Kasik, piluk,* and *rujuk*, are adjectival
particles, denoting respectively, folly, meanness, and depreciation,
as *Innuk*, a man; *Innukasik*, a foolish man; *Innukjiluk*, a mean man;

and *Innurujuk*, a contemptible man. *Pait* is a particle of multitude, as *Ujarak*, a stone; *Ujararpait*, many stones; and *Ujararpagsuit*, a great many stones. *Ngajak* is a particle denoting mixture, as *Kablunak*, a Dane; *Kablunañgajak*, a half-caste. *Ták* and *Tokák* are respectively old and new, as *Anórak*, clothes; *Anorartak*, new clothes; *Anorartokak*, old clothes. *Mio* means an inhabitant, as *Narsak*, a valley; *Narsarmio*, a dweller in the valley. *Minek* is a piece of anything, as *Kissuk*, wood; *Kissumiinek*, a piece of wood. *Nek* is a participle termination, as *Sinigpok*, a sleeper; *Sinignek*, sleeping; while *Fik* forms a noun, as *Igsiavhok*, a sitter; *Igsiav-fik*, a stool. *Usek* has a similar office, as *Okarpok*, a speaker; *Okar-usek*, a word.

The personal pronouns are :—

Uanga,	I.	*Illipsi,*	ye.
Uagut,	we.	*Oma,*	he.
Illit,	thou.	*Okkoa,*	they.

The possessive pronouns are formed by the addition of particles to the root, as :—

Nuna, land.

Nunaga,	my land.	*Nunarse,*	your land.
Nunarput,	our land.	*Nuná Nunanga,*	his land.
Nunat,	thy land.	*Nunartik,*	their land.

Iglu, house.

Igluga,	my house.	*Iglua,*	his house.
Iglut,	thy house.	*Iglutit,*	thy house.

When the signification is transitive, passing from one to another, the pronoun is declined differently, the endings being *ama*, my; *auit*, thy. As *Nalegak*, a chief; *Nalegama*, my chief; *Nalegauit*, thy chief does so and so to me, you, or him.

The interrogatives are as follows :—

What,	*suna.*	Which,	*sút.*
When,	*kakugo.*	Who,	*kiná.*
Where,	*sumá.*	Whose,	*kiá.*

The relative pronouns are :—

Those,	*Ieko.*	That,	*ivna.*

THE VERB.

The verbs have been divided into five conjugations, according to their terminations :—

1. *Kpok*	as	*ermikpok,*	he washes himself.
2. *Kpok*	,,	*matarpok,*	he undresses.
3. *Pok*	,,	*egipok,*	he casts away.
4. *Ok*	,,	*pyok,*	he gets.
5. *Au*	,.	*irsigau.*	he beholds.

The negative goes through every mood and tense of every verb. It is expressed by *ngilak*, as *ermingilak*, he does not wash himself.

The third person singular indicative is the root from whence all the other persons a reformed, by affixing the pronoun, as *Ermikpok*, he washes; *Ermikpotit*, you wash.

There are three tenses, present, preterite, and future. The present is indicated by a *p*; the perfect by a *t* or *s*; and the future, in two forms, *sav* and *goma*; as—

Ermikpok,	he washes.
Ermiksok,	he has washed.
Ermisavok,	he will wash.
Ermigomarpok,	he will wash sometime hence.

The moods are six in number. The indicative in *kpok*; the interrogative in *kpa*; the imperative in two forms, one persuasive in *na*, the other more imperative in *git*; the permissive also in two forms, one exacting, the other requesting, in *gle* and *naunga*; the causal in *kame*; the conditional in *kune*; and the infinitive in three forms. As :—

Indicative.	*Ermikpok,*	he washes.
Interrogative.	*Ermikpa,*	does he wash ?
Imperative.	1. *Ermina,*	please to wash.
	2. *Ermigit,*	wash.
Permissive.	1. *Ermigle,*	let me wash.
	2. *Erminaunga,*	„ „
Causal.	*Ermikame,*	because he has washed.
Conditional.	*Ermikune,*	if he washes.
Infinitive.	*Ermiklune,*	to wash.

Takurâ.	he sees or saw him.
Takuronk,	did he see him ?
Takuliuk,	may he see him ?
Takungmago,	because he saw him.
Takugpago,	when he saw him.
Takuedlugo,	seeing.
Takuyâ,	that he saw him.

The verb *pyok*, to do or get, is used in many cases in conjunction with the infinitive of other verbs.

The present indicative of the active verb is thus conjugated :—

He travels,	*autdlarpok.*		He knows,	*ilipok.*
You travel,	*autdlarpatit.*		You know,	*ilipatit.*
I travel,	*autdlarpunga.*		I know,	*ilipunga.*
They travel,	*autdlarput.*		They know,	*iliput.*
Ye travel,	*autdlarpusé.*		Ye know,	*ilipusé.*
We travel,	*autdlarpugut.*		We know,	*ilipugut.*

He loves,	*asavok.*		He sleeps,	*siningpok.*
You love,	*asavutit.*		You sleep,	*siningputit.*
I love,	*asarunga.*		I sleep,	*siningpunga.*
They love,	*asarut.*		They suffer,	*mikkaut.*
Ye love,	*asarusé.*		Ye suffer,	*mikkause.*
We love,	*asarugut.*		We suffer,	*mikkaugut.*

He comes,	*aggerpok.*	They come,	*aggerput.*
You come,	*aggerputit.*	Ye come,	*aggerpusé.*
I come,	*aggerpunga.*	We come,	*aggerpugut.*

The conjugations, through all moods and tenses, are effected by the use of the personal pronouns; and there are transitions, when the action passes from one person to another, as in several American languages : as—

He washes himself,	*ermikpok.*
You wash yourself,	*ermikputit.*
I wash myself,	*ermikpunga.*
They wash themselves,	*ermikput.*
They two wash themselves,	*ermikpuk.*
Ye wash yourselves,	*ermikpusé.*
We wash ourselves,	*ermikpugut.*
We two wash ourselves,	*ermikpuguk.*

Every mood and tense is thus inflected with the suffixes of the persons, ringing the changes in each transition; as, he washes himself, he washes you, he washes them, he washes us; and so with all the other persons. For example, to conjugate through all the persons washing a third person we have—

He washes him,	*ermikpa.*	They wash him,	*ermikpat.*
You wash him,	*ermikpet.*	Ye wash him,	*ermikparse.*
I wash him,	*ermikpara.*	We wash him,	*ermikparput.*

He sees it,	*takuvâ.*	He sees me,	*takuvânga.*
You see it,	*takuvat.*	You see her,	*takuvatit.*
I see it,	*takuvara.*		

| The fox sees it, | *teriaviak takuvâ.* |
| The fox saw him, | *teriaviap takuvâ.* |

The participle, which supplies the place of an adjective, is the same as the preterite, *Ermiksok*, washed. The future is *Ermissirsok*, he will wash.

The principal auxiliary verb is *pyok;* with which, or with various particles, an infinity of words are formed into one, the last only being conjugated. Crantz gives an instance of this, where a single word expresses what in English requires seventeen :—" He says that you also will go away quickly in like manner and buy a pretty knife." In Eskimo this is—*Savigiksiniariartokasuaromaryotittogog*—composed as follows :—

Savig,	a knife.	*Omar,*	wilt.
Ik,	pretty.	*Y,*	in like manner.
Sini,	buy.	*Otit,*	thou.
Ariartok,	go away.	*Tog,*	also.
Asuar,	hasten.	*Og,*	he says.

The Eskimo language is peculiar in the use of the affirmative and negative conjunctions, *ap* and *nagga*. To the question, *Pioma-ngilatit*, " Wilt thou not have this?" if the questioned person will have it, he must answer, *nagga*, " No." If he will not have it, he must say, *ap*, " Yes,"—*piomangilanga*, " I will not have it."

VOCABULARIES.

MAMMALS.

Cetaceans.

Male, *angut.* Female, *arnat.*

Whale,	*maglagdlit, arvek.*	Fin fish,	*tunulik.*
Sperm whale,	*kegutilik.*	Little finner	*tikagutlik.*
Bottle-nose,	*kiporkak, nisaruak, anarnak.*	Porpoise,	*piglutok.*
		Grampus,	*ardluik.*
White whale,	*kilalkak.*	"	*ardluarsuk.*
Narwhal,	*tugalik, kernertok.*		

Seals.

Walrus	*aurek, aruek.*	Harp seal,	*atak, atarsuak, atarneit-suak.*
Bearded seal,	*ngsuk, usuk.*		
Hooded seal,	*kakortak, nátsiersuak.*	Hispid seal,	*natsek, natsidlak.*

Parts of Whales and Seals, &c.

Whale bone,	*sorkak.*	Seal's hind-flipper,	*okpotik.*
Blubber,	*ossnk.*	Ivory,	*saunek.*
Walrus tusk,	*togak.*	Seal hole,	*atluk (aglo?)*
Seal's fore-flipper,	*tellerok tallik.*		

Carnivora.

Bear,	*nanuk.*	Blue fox,	*terianiak kernetok.*
Wolf,	*amaruk.*	White fox,	*terianiak kakortak.*
Fox,	*teriuniak.*	Dog,	*kingmek.*

Ruminants.

Reindeer,	*tuktu.*	Young deer,	*noraitsok.*
Great reindeer,	*angisok, panguek, nugatugak.*	Fawn,	*norkak.*
		Musk ox,	*uming-mak.*
Doe,	*kulavak.*		

Rodents.

Hare,	*ukalek, tulukat.*	Mouse,	*teriuk.*
Rat,	*kitsuk.*		

Parts of Reindeer.

Antler,	*naksuk, agiak.*		Venison,	*nekké.*
Fur,	*mituk, illupakot.*		Liver,	*tinguk.*
Tail,	*pamiok.*			

BIRDS.

Birds of Prey.

Bird,	*tingmiak.*		Great snowy owl,	*ugpik* or	*opik,*
Cinereous eagle,	*nugtoralik.*			*opiksoak.*	
Gyr falcon,	*kingsariarsuk, kigavik.*		Raven,	*tulueak.*	

Divers and Guillemots.

Great auk,	*isarukitsok.*		Loom,	*agpa.*
Cormorant,	*okaitsok.*		Dovekey,	*serfak.*
Great northern diver,	*tugdlik.*		Little auk,	*agpaliarsuk.*
Red-throated diver,	*karsak.*			

Gulls.

Glaucous gull,	*nayak.*		Ivory gull,	*najuarssuk.*
Kittiwake,	*talerak.*		Skua,	*isurgak.*
Fulmar petrel,	*kakugdluk.*		Tern,	*imerkutailak.*

Ducks and Geese.

Swan,	*kugsuk.*		Eider duck,	*metek.*
Brent goose,	*nerdlek.*		Long-tailed duck,	*agdlek.*
King duck,	*kingalik.*		Harlequin duck,	*tornariasuk.*

Ptarmigan, Snipe, &c.

Ptarmigan,	*akigsek.*		Sandpiper,	*kayungoak.*
Plover,	*kajurdlak.*		,,	*tugagvorjok.*
Snipe,	*taluifak.*		,,	*nalumasortok.*
Phalarope,	*sarforsuk.*			

Finches, &c.

Snow bunting,	*korpanuk.*		Snow bunting,	*tirgivok, agdlorpok,*
,, ,,	*korpaluarsok.*			*mitsimaput.*
,, ,,	*kopurnarsuk.*		,, ,,	*mipok, pikiarpok,*
,, ,,	*kapiarak.*			*iravok.*

Parts of Birds.

Quill,	*sullok.*		Beak,	*sigguk.*
Feather,	*initkut.*		Talon,	*isiguk.*
Wing,	*isarkok.*		Egg,	*maunik.*
Tail,	*pappik.*		Nest,	*uvlo.*

P

FISH.

Fish,	*akalua.*	Bull head,	*kaniok.*
Salmon,	*erkaluit.*	Red char,	*ekelluk.*
Capelin,	*augmaksak.*	Herring,	*augmaset.*
Char,	*aulisogik.*	„	*sarawdlit.*
Shark,	*ekellurksoak.*	„	*ogak.*
Hallibut,	*kalleraglik.*	Eel,	*puttorotok.*
Perch,	*sullupaugak.*		

CRUSTACEA, MOLLUSCS.

Shrimp,	*kinguk.*	Snail,	*sinterok.*
Crab,	*aksegiak.*	Mussel,	*sigguk.*
Louse,	*komak.*	Oyster,	*kiksauausak.*
Worm,	*kuglumiak.*	Star fish,	*aksogiarsoit.*

INSECTS.

Fly,	*niviugak.*	Flea,	*pigleitok.*
Mosquito,	*ippernak.*	Butterfly,	*tarku-tarkisak.*
Bee,	*egitsak.*	Spider,	*ausiek.*

PLANTS.

Wood,	*kissuk (kresut* of Kane).	Sorrel,	*kupeklusuit.*
Tree,	*orpik.*	Saxifrage,	*kakilarnekot.*
Bush,	*neturkok.*	Cress,	*iriykut.*
Dwarf willow,	*sersåt, nunaugiuit.*	Club moss,	*irsutit.*
Dwarf birch,	*avalakissat.*	Moss,	*maunek, tinganset.*
Crowberry,	*paurnak.*	Lichen,	*kod yutit, ti-rauyat, o-ku-*
Bilberry,	*kigulernek.*		*yut.*
Whortleberry,	*kingnernat.*	Root,	*sordlait, sortlak.*
Scurvy grass,	*kungordlet.*	Bark,	*kallipek.*
Grass,	*irik.*	Leaf,	*karké, nuni-verset.*
Reed,	*ivuikot.*		

MEN AND SPIRITS.

Human being,	*innuk.*	A spirit, familiar,	*torngak.*
Mankind,	*innuit.*	Inhabitant of the air,	*innerterriosok.*
Man of Greenland,	*karalit.*	Evil spirit of the air,	*erloersortok.*
Man not of Greenland,	*innuit tekornartet.*	Sea spirits,	*kongensetokit.*
Dane,	*Kabluuah.*	Fire spirits,	*inguersoit.*
Englishman,	*Tulluk.*	Mountain spirit,	*tunuersoit.*
Dutchman,	*Arseniak.*	Evil spirits,	*erkiglit.*
Chief,	*nalegak.*	Genius of the winds,	*sillagiksartok.*
Priest ("He is very	*angekok.*	Genius of food,	*nerrim-innuet.*
great"),		Ghost,	*auersak, angiak.*
Wizard,	*issuitok, illioitsok.*	The famous wise one,	*angekut poglit.*
"He that is above,"	*pirksomu* (Egede).	Heaven,	*killak.*
Good spirit,	*torngarsuk* (Crantz,	Hell,	*tornarsivik.*
	i. p. 206).	A tribunal of redress,	*innapok.*

HUMAN BEINGS.

Man,	*anguk.*	Children	*merdlukit.*
Woman,	*arnak.*	Baby,	*naulungiak.*
Boy.	*nukakpiak.*	Old man,	*ittok, augutokak.*
Girl,	*niviarsiak.*	Old woman,	*arnatokak.*
Child,	*meruk, kittornak.*		

RELATIONSHIPS, CEREMONIES, &c.

Ancestors,	*angejokait.*	Nephew,	*katteng-utiksiak.*
Father,	*atatak.*	Friend,	*illet, ikingut.*
Mother,	*annanak.*	Companion,	*tessiorpok.*
Father or mother-in-law,	*sekke.*	Guide,	*tessiosiok.*
Husband,	*uria, uinga.*	Custom,	*illerkok.*
Wife,	*nuklia, nulli- anga.*	Song,	*innernek.*
		Dance,	*ketingnek.*
Son,	*erninga.*	Feast,	*nerkanek.*
Daughter,	*pannik.*	Birth,	*innung-ornek.*
Brother,	*kattangut.*	Pregnant,	*nartavok.*
Sister,	*kattangut-ar- nak.*	Life,	*innunek.*
		Death,	*toko.*
Uncle,	*akkek, angak.*	Corpse,	*tokossok.*
Aunt,	*ayak, aitsak.*	Grave,	*illivik.*

PARTS OF THE BODY.

Body,	*time.*	Wrist,	*pavsik.*
Skin, *Innúb amid?* (W.)	*amek.*	Hand,	*aksait.*
Head,	*niakok.*	Left hand,	*sacemik.*
Brains,	*karresak.*	Right hand,	*tellerpik.*
Forehead,	*ka-nk.*	Knuckle,	*kangmak.*
Face,	*kenak.*	Forefinger,	*tikvk.*
Hair,	*nutsak.*	Middle finger,	*keterlek.*
Beard,	*umik, ersarutit.*	Little finger,	*erketkok.*
Eye (*Irsé,* Hans Egede),	*isé.*	Thumb,	*kublo.*
Eyelash,	*kemeriak.*	Nail,	*kukkik.*
Tears,	*koblit.*	Liver,	*tinguk.*
Nose,	*kingak.*	Waist,	*timmé.*
Mouth,	*kanek.*	Belly,	*nak, neksek.*
Teeth,	*kigutit.*	Stomach,	*akyorok.*
Tongue,	*okak.*	Navel,	*kallasek.*
Gums,	*itkitka.*	Bowels,	*innello-lut.*
Breath,	*annernek.*	Buttocks,	*nullok.*
Lip,	*kartlo.*	Kidney,	*tarto.*
Cheek,	*ulluak.*	Thigh,	*ukpet, koktorak.*
Chin,	*taplo.*	Hip,	*siciak.*
Ear,	*siut.*	Knee,	*siskok, serkok.*
Neck,	*kongesek.*	Leg,	*kannah* (Hans
Windpipe,	*kingak, tortluk.*		Egede), *niö.*
Shoulder,	*tué.*	Ankle,	*nappasortak, kanginak.*
Back,	*kaltigok.*		
Backbone,	*kemerluk.*	Foot,	*isiket.*
Side,	*sennarak.*	Step,	*abloruek.*
Breast,	*sekkiok.*	Foot print,	*tunmé.*
Lungs,	*puak.*	Heel,	*kimik.*
Heart,	*umat.*	Toe,	*puttoguk.*
Bosom,	*iciengek.*	Blood,	*auk.*
Milk,	*innuk.*	Vein,	*tarkuk.*
Arm (right),	*tellerk.*	Bone,	*saunek.*
„ above the elbow,	*aksaut.*	Wound	*iké.*
„ below the elbow,	*aksarkok.*	Bruise	*ajuak.*
Elbow,	*ikusik.*		

House and Furniture.

House,	*iglu.*	Lamp,	*kotluk, kollek.*
Winter entrance,	*tossut.*	Water-kid,	*imertaut.*
Door,	*matto.*	Cooking-pot,	*kolupsut.*
Window,	*igaluk.*	Caldron,	*utak* (Kane).
Seat,	*igsiavik.*	Cup,	*imertarfik.*
Bed,	*sinik-rik, sinik-tarpik.*	Jar,	*kongmuvetak.*
		Ladle,	*alluksuut.*
Blanket,	*kepiksoak.*	Food,	*nerriseksuk.*
Stove,	*kissarsuut.*	Soup,	*kuyok.*
Smoke,	*puyok, iseriek.*	Broom,	*senniut.*

Tent.

Tent,	*tupek.*	Tent pole,	*kannak.*
Hoop for tent,	*tupekarfiksuk.*		

Clothes.

Clothes,	*auorak, auorakset.*	Boot,	*kamik.*
Dressed leather,	*amersoak.*	Stocking,	*allersé* (Crantz).
Fox-skin jumper,	*kapetah.*	Gaiter,	*suiget.*
Under jacket,	*attigek.*	Band for a woman's hair,	*kelerutik.*
Hood,	*nessak.*		
Cloak,	*kavoyak, uliksoak.*	Shirt,	*illupak.*
Mitten,	*arkit merkusalik, aket.*	Skin shirt,	*metnérok.*
		Beads,	*sapang-at, tuglautit.*
Breeches,	*karlik, nannuk.*	Ring,	*akoangmio.*
Button,	*athet.*		

Implements.

Sledge,	*kammutik.*	Stone arrow-head,	*karkovit, ullugsak, nyarak.*
Dog harness,	*annut.*		
Whip,	*ipperautik.*	Dart for birds,	*nugpit.*
Trap,	*pullet. tutdet.*	Three-pronged dart,	*egeniak.*
Canoe,	*kuyak.*	Knife,	*sarik.*
Woman's boat,	*umiak.*	„	*pillautak.*
Paddle,	*pautik, eput.*	„	*okutartok.*
Harpoon,	*unak, muligeit.*	Stone knife,	*pillektavit.*
Seal-bladder float,	*avealetok.*	Woman's knife,	*ulli.*
Moveable lance-head,	*nuagkak.*	Adze,	*ullimaut, senningasok.*
Spear for large seal,	*agligalik.*		
Spear for small seal,	*agliak.*	Saw,	*pillektok.*
Whale spear,	*kallugiah.*	Drill,	*keolerut.*
Deer spear,	*narkok, karsok.*	Pricker,	*tók toua.*
Line of seal-skin,	*alluuak.*	Needle,	*mitkut.*
Harpoon-line,	*alak ardlet.*	Thimble,	*tikek.*
Rope,	*aklunak (?).*	Fish-hook,	*okoumersak, kar-sakuk.*
Bow,	*pisiksek.*		
Arrow,	*karsok.*	Bone instrument for discovering seals under the ice,	*sekko, saumeruit.*
Quiver and bow-case,	*póik, pisiksek.*		
Bowstring,	*narkukté.*		

TIMES AND SEASONS.

Time,	*nelliunek.*	Spring,	*uperuik.*
Day,	*uvtlok.*	Autumn,	*ukiak.*
Daylight,	*kaumanek.*	Winter,	*ukipok, ukiok.*
Night,	*unuak.*	Weather,	*silla.*
Evening,	*unuk.*	Sky (cloudy),	*nuia.*
Midnight,	*unuk, ketka.*	Sky,	*killek.*
Morning,	*ublumi, utlukut.*	Fog,	*puyorpok.*
Summer,	*ausak.*		

WEATHER.

Sun,	*sekkinnék.*	Aurora Borealis,	*arksuonerit.*
Moon,	*kaumot, anningat.*	Rain,	*sialluk.*
Sunrise,	*sekkernub-pullinek.*	Snow,	*kannik.*
Sunset,	*sekkernub-larksetsek.*	Dew,	*isugutanek.*
Eclipse,	*sekkernub.*	Rainbow,	*merinuak, pusisak.*
Shadow,	*tarkak.*	Lightning,	*inguek-perlenguek.*
Star,	*ubloriak.*	It thunders,	*katlekpok.*

WIND.

Wind,	*annoré.*	North-west wind,	*tamaké.*
North wind,	*aranguak, awanguek.*	South-east wind,	*numisarnek.*
South wind,	*kienguk.*	North-east wind,	*awangnusīk.*
East wind,	*agsarnek.*	It is calm,	*kaitsorpok.*
West wind,	*kwanguak.*	Open air,	*sillamepok.*
South-west wind,	*kigek.*		

ELEMENTS.

Earth,	*nuna.*	Flame,	*iknellanek.*
Air,	*silla.*	Water,	*imak.*
Fire,	*inguek.*		

LAND.

Boundary,	*kigdlik.*	Chalk,	*aglaut, kakortok.*
Mountain,	*kakak.*	Salt,	*tarruitsut.*
Valley, level land,	*narsak.*	Quartz,	*usugiak.*
Island,	*kikertak.*	Felspar,	*angmak.*
Cape,	*kangek.*	Granite,	*aunnaussat.*
Coast,	*siksak.*	Coal,	*aumarutikset, aumarsuit.*
Cliff,	*imnak, innarsak.*		
Reef of rock,	*ikarlok.*	Limestone,	*kerchok, kerchursiak.*
Shoal,	*siorarsoak.*		
Rock,	*kiugiktok.*	Iron,	*savik.*
Stone,	*uyarak.*	Cryolite,	*aussap-ujará.*
Heap,	*peksoit.*	Graphite,	*akerdlusak.*
Hole,	*kidluk.*	Steatite,	*urkusigssak.*
Cave,	*karosuk.*	Flint,	*aut-dlait.*
Gravel,	*siorak.*	Schist,	*kalekutigsiat.*
Clay,	*markhak.*	Sand,	*siorket, siorak.*

ICE.

Snow,	*kannik, apút.*	Ice,	*sikko.*
Snow-drift,	*aaniyo.*	Iceberg,	*pikuluyak, illuliak.*
Hummock,	*kagiealwaluk.*	Glacier,	*sernik-suak.*

200 VOCABULARIES.

SEA.

Sea,	*imarpíksoak, innaksoak.*	Ebb-tide,	*tinné.*
Salt-water,	*imak.*	Flood-tide,	*ullé.*
Wave,	*malik.*		

WATER.

Water,	*imak, ermit.*	River,	*kouk, kouksoak.*
Bay, lake,	*kangerluk.*	Pond,	*tasek.*
Fiord, inlet,	*kangerlersuk, tessek.*	Strait,	*ikarasek.*
Stream,	*sarwak, koyeisiak.*	Ford,	*ikarut.*

COLOURS.

White,	*kakortok.*	Red,	*aukpallartok.*
Black,	*kernerpok.*	Yellow,	*korsorpallukpok.*
Brown,	*kayoapok.*	Green,	*tungyoktok.*
Blue,	*tung-iyorpok.*	Grey,	*kernaluartok.*

NUMERALS.

1,	*atausek.*	8,	*arwenguet-pingasut.*
2,	*marluk.*	9,	*kolingiluet.*
3,	*pingasut.*	10,	*kolit.*
4,	*sisamet.*	First,	*siudlek.*
5,	*tedlimet.*	Second,	*atla.*
6,	*arwenguet-orpenek.*	Half,	*auwiktok, keterkot.*
7,	*arwenguet-marluk.*	All,	*tamat.*

SENSATION.

Cold,	*dikpok, keyulerpok.*	Warm,	*kiek.*
Frost-bite,	*kenaersiuek.*	Hot,	*kiekpok.*
Frozen,	*kerivok.*		

SIZE AND FORM.

Great,	*angisok.*	Thin,	*amitsok, sellusok.*
Little,	*mikisok.*	Flat,	*saitok.*
Broad,	*sillikpok.*	Round,	*augmalortok.*
Narrow,	*amitsok.*	Square,	*terkerkolik, to-artok.*
Fat,	*puellawok, kuiuiwok.*	All,	*tamarmik.*

TASTE.

Sweet,	*mamarkauk.*	Bitter,	*kassilipok.*
Sour,	*sernarpok.*		

ADJECTIVES.

Blind,	*takpepok.*	Handsome,	*pinnersok.*
Deaf,	*tusilarpok, tusianipok.*	Pretty,	*innekonak.*
Dumb,	*okauitsok.*	Strong,	*nekoarpok. pisuk.*
Be silent!	*nipang-erniarit.*	Weak,	*kayeryuartok.*
Beauty,	*pinnersusek.*	True	*ilunnt.*
Very beautiful,	*pirivigpok.*		

ADVERBS OF PLACE.

Here,	*ma.*	East,	*pac.*
There,	*tas.*	West,	*som.*
North or right (looking seaward),	*}ac.*	Above (landwards),	*pik.*
		Senwards or west,	*iktorpok.*
South or left (looking seaward),	*}kar.*	South, where the sun goes,	*kig.*
Where,	*sume.*	Within,	*kum.*
Whence,	*sumit.*		

VERBS (*alphabetical*).

He is able,	*pikkorik-pok, sinnawok.*	He counts,	*kissipok.*
		He is courageous,	*erksing-iluk.*
He accepts,	*tiguwok.*	He covers,	*talitserpok.*
He accompanies,	*aypara-ok.*	He is cowardly,	*iktorpok.*
He is alive,	*inguwok, umasok.*	He creeps,	*pauung-orpok.*
He alters,	*adlangortipok.*	He crosses over,	*ikurpok.*
He is angry,	*niugekpok.*	He is cruel,	*unniar-titsiok, erksinartok.*
He answers,	*akkiniarpok.*		
He approaches,	*kaunining-orpok.*	He cries,	*kiyarok.*
He is arrived,	*tikipok.*	He cuts,	*kippiluk.*
He asks,	*aperaok, apersorpok.*	It is damp,	*kauserpok.*
He is awake,	*iterpok, erkomawok.*	He dances,	*ketinguek.*
He is away,	*tamuk, aularpok.*	He dares,	*iktungilak.*
He is bad,	*ajorpok.*	It is dangerous,	*nauričnarpok.*
He is bald,	*niyakang-ilak.*	It is dark,	*tarpok.*
She is bashful,	*kangusak-pok, tessitsiok.*	He is dead,	*tokorok.*
		It is deep,	*itinok.*
He beats,	*unatarpok.*	He defends,	*sernigauk.*
He begs,	*kennwok, tuksiapok.*	He denies,	*uissiarpok.*
He beholds,	*irsigau.*	He destroys,	*asserorpok.*
He believes,	*operpok, isamawok.*	He dies,	*tokowok.*
He bellows,	*miagorpok.*	It is difficult,	*ayornarpok.*
It bends,	*perikpok.*	He is dirty,	*ippertok.*
It blazes,	*ikuellapok.*	It is distant,	*ung-esikpok.*
It bleeds,	*aunarpok.*	He disputes,	*aksortriok.*
It blocks up,	*illuwok.*	He dives,	*akarpok.*
He blows,	*supporpok.*	He dives as a seal,	*pullavok.*
He blows (as a whale),	*aunersarpok.*	He divides,	*avipok.*
		He doctors,	*nakkursak.*
It boils,	*kallapok.*	He does,	*piyok, illiorpok.*
He is born,	*inun-simawok.*	He drags,	*uniarpok.*
He bows,	*sikkipok.*	He dreams,	*sinektorpok.*
He brings,	*apok.*	He dresses,	*nimerpok.*
It is broken,	*nappiwok.*	He drinks,	*imerpok.*
It burns,	*ikumawok.*	He drives,	*kemukserpok.*
He buys,	*pisiniarpok.*	It drops,	*kusserpok.*
He carries,	*nungmakpok.*	He drowns,	*epicok.*
He casts away,	*egipok.*	He is drunk,	*pattorkawok.*
He cheats,	*milekpok.*	It is dry,	*pannerpok.*
He is cheerful,	*unennarpok.*	He eats,	*nerriok.*
He chews,	*tamorpok.*	It is empty,	*inatkang-iluk.*
He chooses,	*kennerpok.*	He embarks,	*ikiwok.*
He chops,	*serkoniupok.*	He exchanges,	*tauserpok.*
He climbs,	*kakkiwok.*	He faints,	*irkang-uwok.*
He comes,	*aggerpok.*	It falls,	*nukkartok.*
He commands,	*pekowok.*	He is far off,	*ung-esikpok.*
He cooks,	*igawok.*	He makes fast,	*aulegong-ersakpok.*
He coughs,	*koersorpok.*	He is fat,	*pullawok, kuiniwok.*

Verbs—*continued.*

He is fatigued,	*kassuvok, nungavok.*	He lifts,	*kicikpok.*
He is feeble,	*nukingiarpok.*	He is living,	*umavok.*
He feeds,	*nerrisipok.*	He looks,	*tekovok.*
He feels,	*serptikpok.*	He loosens,	*porpok, pellukpok.*
He fetches,	*aiokpok.*	He loves,	*asavok.*
They are few,	*ikitput.*	She loves him.	*assavá.*
He finds,	*nenniuk.*	Married,	*nullialik.*
It is finished,	*euerpok.*	He meets him,	*napipok.*
He fishes,	*aulisarpok.*	He is merry,	*kemavok.*
He flies,	*kemavok.*	It melts,	*aupok.*
It floats,	*puktavok.*	He mourns,	*alliyi-sukpok.*
He follows,	*mullikpa.*	He murders,	*innuorpok.*
He forbids,	*innert-itpok.*	He is naked,	*mattang-asuk.*
He forgives,	*salmaupok.*	He nods,	*angerpok*
It freezes,	*kerivok.*	He obeys,	*natekpok.*
He gets,	*pyok.*	He is old,	*utokavok.*
Give me,	*tunning-a.*	It oversets,	*kinguvok.*
He is glad,	*tipsi-sukpok.*	He paddles,	*pautakpok.*
He gnaws,	*mikarsigauk.*	He pants,	*anuersertarpok.*
He goes,	*pisugpok.*	He plucks off,	*nnuiokpok.*
Go down!	*atterpok.*	He plugs up,	*siniksok.*
Go in!	*iserpok.*	He pours,	*koisok.*
Go out!	*annipok.*	He pricks,	*kupivok.*
He is gone,	*aularpok.*	He pulls,	*onarpok.*
I am good,	*ayung-ilanga.*	He pulls one's hair,	*nutsukpok.*
He is great,	*angisok.*	He punishes,	*pittarpok.*
It grazes (deer),	*viktorpok.*	He purchases,	*pisiniarpok.*
He is greedy,	*nerkersok.*	He pushes,	*ayekpok.*
He groans,	*anner-saumivok.*	It is putrid,	*tipitovok.*
It grows,	*nauvok.*	He is quiet,	*allianurpok.*
He hangs,	*niving-arpok.*	It rains,	*sial-lukpok.*
It is hard,	*arktornarpok.*	He raises,	*koblarpok.*
He made haste,	*piesuarpok.*	He receives!	*piok, tignvok.*
He hears,	*tussarpok.*	Remember!	*erkai-niarit.*
He is healthy,	*ke avok, atsuilivok.*	He returns,	*utertok.*
It is heavy,	*okenwipok, arktorner-pok.*	That is right,	*illnarpok.*
		It rises (as the tide),	*ullitsar-torpok.*
He is tall,	*angisoak.*		
It is high,	*porturok.*	He is a rogue,	*tigliktok.*
Hold fast,	*tiguk.*	It rolls,	*aksakartok.*
He is honest,	*pelkoser-suitpok.*	It is rough,	*manning-itsok.*
He hopes,	*nerrikupok.*	It is round,	*angmalortok.*
How do you do?	*kannong-illettit?*	He rubs,	*aggiarpok.*
He is hungry,	*pertlillerpok*	He runs,	*arkpatok.*
He is idle,	*innalyn-ivya.*	He is sad,	*alleyn-orpok.*
He jumps,	*pissikpok.*	He saves,	*aulatpa.*
He kicks,	*tungnarpok.*	He says,	*okarpok.*
He kills,	*tokopok.*	He scratches,	*ketsukpok, kuniptok.*
He is kind,	*innuksi-arnerpok.*	He screams,	*tortlorpok, nippanotit.*
He kisses,	*kunnikpok.*	He is gone seal-ing,	*auguni-arpok, arkö-isok.*
He kneels,	*siskomiartok.*		
He ties a knot,	*kebrsorpok.*	He sees him,	*takuvá.*
He unties a knot,	*kebrusuerpok.*	He sells it,	*pissiniutiga.*
He knows,	*ilipok.*	He sends,	*neksiupa.*
He laughs,	*iglarpok.*	She sews,	*suersortok.*
He leads,	*tessiorpok.*	It shines,	*keblerpok.*
He learns,	*ilitpok.*	He shovels,	*nivagpok.*
He lends,	*attartorpok.*	He shouts,	*tortlu-lortok.*
He licks,	*alluktorpok.*	He is sick,	*nappurpok.*
He lies down,	*nellersok.*	He sighs,	*annersartopok.*

VERBS—*continued.*

He sings,	*illerkorsorpok, inner-*	He swallows,	*iveok.*
	tok.	He sweats,	*ailarpok,*
It sinks,	*kiveisok.*	It swells,	*poveitok.*
Sit down,	*ingitit.*	He swims,	*nelluktok.*
He slaps,	*unatarpok.*	Take away !	*pi-uk !*
He sleeps,	*siningpok.*[1]	Take care !	*miauersorpit.*
It slides down,	*sissusok.*	Take it !	*tignik.*
He is slow,	*ingarleng-it*ok.*	He talks,	*okalluktok.*
He is small,	*mikisung-uieok.*	He teaches,	*ayorker-sakpa.*
He smells,	*namaruk.*	He tears,	*alliktorpok.*
He smiles,	*kongiularpok.*	I am thirsty,	*imeruktunga, killaler-*
It smokes,	*pu-yortok.*		*pok.*
It is smooth,	*maniqpok.*	It thaws,	*aunatok* (Kane).
He sneezes,	*tagioktok.*	He throws a spear,	*nuulikpok.*
He snores,	*sinikpullukpok.*	He tickles,	*koluekpok.*
It snows,	*kaanerpok, niptarpok.*	He ties,	*kerlerpok.*
It is soft,	*ketuktok.*	I am tired,	*kassuieonga.*
It is sore,	*annernartok.*	He travels,	*ingerleook-autdlarpok.*
He speaks,	*okalluktok.*	He trembles,	*saynkpok.*
He spits,	*kesersok.*	He twists,	*kaipsiok.*
He splits it,	*koppisok.*	He undresses,	*maturpok.*
He squeezes,	*erkiterpok.*	He vomits,	*meriarpok.*
He stabs,	*kappisok.*	He walks,	*pissuktok.*
He steals,	*tiglikpok.*	He washes,	*ernikpok.*
Stop !	*unikpit, tessa !*	He watches,	*pigorpok.*
He stretches,	*isuilukpok.*	He is well,	*somanglluk.*
He strikes him,	*unartarpa, tigluksak-*	Well done !	*ayung-illuk.*
	pa.	He went,	*peartorpok.*
He is strong,	*nekoarpok.*	It is wet,	*kausersok.*
She suckles,	*miluk-titsiok.*	He whips,	*ipperartorpâ.*
He suffers,	*mikekuwk.*	He whistles,	*ningiartok.*
He is surprised,	*tettamiok, tupigosuk-*	He is young,	*innuo-suktok.*
	pok.		

[1] Washington has *sinniktok.*

LIST OF NAMES OF PLACES IN GREENLAND.

Eskimo.	Meaning.	Norman.	Danish.	Admiralty Chart and Latitude.	Remarks.
			EAST COAST OF GREENLAND.		
				Cape Barclay of Scoresby (1822) 69° 13' N.	[*Between C. Barclay and Danels Öer, 223 miles undiscovered.*]
			Dauels Öer of Danel, 1652	63° 30' N.	Gunbiörn Skerries, according to Graah. Sighted by David Danel in 1653.
			C. Thyge Bruhe.		
			Egedes og Rothes F.	65° 15' 36" N.	
			Dannebrogs Öe	? (65° 18' N.)	
			C. Holm	63° 15' N.	
			C. Guldbrand.		
			Venloan Island	65° 14' N.	"Turn Back" Island of Graah.
			Vahl Island	65° 10' N.	
			Hornemann Island	65° 10' N.	
			Örsted Island	65° 5' N.	
			Cabodo Kong Friderik III.		
			Hvidsadel		
			Ronar's Island	64° 59' N.	
			Sneesdorff Island	64° 58' N.	
			Schnancher Island		
			Kiege Bugt.	64° 55' N.	Graves and huts.
			Skrams Öer		
			C. Torfens.	64° 47' N.	
			Nordbyres Öer.		
			Oxsford F.	64° 35' N.	

Native name	Meaning	European name	Latitude	Remarks
Pœrusermiut		C. Lowenoru	64° 30' N.	
Puisortut ..	"Emerging from the water,"	.. ..	64° 28' N.	
Onœnik	"Landing-place "	Gabels Island ..	64° 22' N.	130 inhabitants in 1829.
Aluik (130 natives)		Gyldenloves F.	64° 18' 50" N.	
Alekajak	Alekak, "elder sister,"			
Tiniartalik	"Having birds"	Krumpens F.	64° 10' N.	20 inhabitants in 1829.
Kekertarsuak	"Large Island" ..	Colberger Heide Glacier	63° 57' N.	
Ukusiksak Island	"Soap-stone"	.. ..	63° 43' N.	
Taterat	"Kittiwakes"	.. ..	65° 40' N.	
Kekertanguak ..	"Small islands" ..	C. Mosting	63° 36' 50" N.	90 inhabitants in 1829.
Kaugvrllugesuak	"Large bay"	Bernstorff's Fiord ..	63° 36' 50" N.	
Ikatamiut Point				
Kenisak Island (90 natives)				
Kekertarssuak Island (75 natives)				
Kangek Point	Cape	C. Moltke	63° 31' N.	75 inhabitants in 1829.
Kekertanguak				
Umanarssuk ..	"Shape of a heart,"			
Kupartalik Island	"Ravine."			
Kangersunek River.				
Nukerfik	Place (fik) where a younger sister or brother (nukak) has been lost (er)	Dronning Maries Dal ..	63° 21' 38" N.	Where Graah wintered.
Ekalumguiut ..	Dwelling-place near the salmon.		63° 31' N.	The place which Mr. Gibbs's expedition attempted to reach. This was the most delightful spot seen by Graah.
		Skioldungen.		
Komok Island ..	Narrow part of inlet			

LIST OF NAMES OF PLACES IN GREENLAND—*continued.*

Eskimo.	Meaning.	Norman.	Danish.	Admiralty Chart and Latitude.	Remarks.
EAST COAST.					
Umanatsalik					
Amarlugtok (20 *natives*)	"Carrying a baby in the hood."	...	...	...	20 inhabitants in 1829.
Amitoarsuk	...	...	Caroline Amalia's Havn	63° 15' N.	
Kutdlernio Point	*Kutdlek,* highest *Mio,* inhabitant.				
Kasiortok	*Kassik,* bad.	...	C. Juel	63° 14' N.	
Asinkassik Island	High promontory connected with the main land by a low isthmus.				
Akuliarusek					
Kongerajck	...		Sehestedt's Fiord	63° 5' N.	
Asiokkusik.					
Takisok (9 *natives*)	Long.		Griffenfeldt's Öe	62° 55' N.	
Umanak	Shape of a heart				
Usunarssak.					
Kekertassetsiak	Rather large island.			62° 46' N.	
Kinarfik Isthmus (14 *natives*)					
Numarssuak	Large land.	...		62° 40' N.	
Tingmiarmiut Mount	*Tingmiak,* bold.				
Asuuit Point				...	
Manitsok Island (20 *natives*)	Rugged	...		62° 30' N.	20 inhabitants in 1829.
Ivellorsiutit	"Search for eiderdown."				
Ekalungmiut.					
Norsak	"Piece of wool."				
Nagtomlik	Eagle.				
Kasingutok.					
Ikernint	Open, unsheltered.				

Name	Meaning	Other name	Position	Remarks
Maligiaset	*Malik*, swell of the sea.		62° 20' N.	
Kangerlingssutsink	Rather large bay	Heinson's F. Ruds Öo..	Heinson's F. Ruds Islands,	
Kekertarssuak	Large island		62° 7' N.	
Kangok	Promontory	C. Stein Bille..	62° 1' N.	
Puisortok Glacier	"Emerging .. from the water."		61° 59' N.	
Kangok		C. Adelaer.		
Sermiput Point	The point of the glacier.		61° 55' N.	
Innugtut	Young people.	C. Rantzau	61° 46' N.	
Asioint.				
Kekertarssuni	Large islands.			
Kekertarssuak	Large island.			
Ukinssortik (50 *natives*)	Wintering-place		61° 31' N.	50 inhabitants in 1829.
Amoritok Fiord	Much wind		61° 32' N.	
Nuk	Point.	C. Tordenskiold	61° 28' N.	
Kunnornek Point				
Kunararuint	Small river.			
Kekertarssuak (75 *natives*)			61° 17' N.	75 inhabitants in 1829.
Anarket Fiord			61° 15' N.	
Tuterat Point (20 *natives*)	Kittiwake	C. Trolle..	61° 10' N.	20 inhabitants in 1829.
Ingiteit Fiord			61° 9' N.	
Kaningeoksusik Point		C. Fischer	61° 5' N.	
Kangarstkassik Point	Bad promontory.		61° 3' N.	
Unamarssuk Island.			61° 2' N.	
Kangerdluluk Fiord	Bad firth.		60° 59' N.	
Kajartalik	"With a kujak"			
Serratuna Point	Magic spells			
Nuk	Point.			
Kujartalik	Having a cleft.			
Unanarsuk (38 *natives*)				38 inhabitants in 1829.
Kangrdlitilmarak.			60° 52' N. (Gmah).	Here the crew of the *Hansa* landed, June 4th, 1870.
Illuluk Island	A peninsula looking like an island.			

LIST OF NAMES OF PLACES IN GREENLAND.

EAST COAST.				
Kangek	Cape	C. Discord	60° 52′ N.	12 inhabitants in 1829.
Ivingmiut (12 *natives*)	*Irik*, "grass"		? (60° 54′ N.), of *Hansa*).	
Ikarsatsiak	A tolerable bay.		61° 0′ N. (*Germans*	
Patursok.			60° 52′ N.	
Kusek	Dripping water	C. Valloe	60° 4′ N.	
Nagtoralik	Eagle		60° 37′ N.	
Norksuk (20 *natives*)		Lindenau Fiord	60° 28′ N.	20 inhabitants in 1829.
Nanitsuk Point	Rather large bay.			
Kangerdlugssuatsiak	Place for fish.			
Kannjorfik				
Kaningæsoknsik	Large point	C. Hvidtfeld	60° 25′ N.	
Nugsuak	"Like a gull's bill".			
Pinguarsuk	Place for whales			
Aricrsiorfik	Sole of foot		Walchendorf's Öe.	60° 12′ N.
Aluk		Korsoe.	Eggers Öe.	60° 9′ N.
Kekertak	Island	Finnsbuda		60° 4′ N. The harbour between Aluk and Prins Christian's Sund.
Ikersarssuak	Large bay	Oltum lengri Fd. (*Longest of all*).	Prins Christians Sund. *So named by Munk in* 1619.	60° 4′ N.
Niakornak	Cape like a head.			
Puisortoarnak.	Large stones.			
Ujararssuit		Skagefjord	Kangia	N.E. of Cape Farewell.
Kangerdluk				
Ujavarssuak.				

Sketch Chart of the
SOUTH COAST
of
GREENLAND
from the Danish Admiralty Survey,
corrected to 1857.

Name	Meaning	Other name		Position	Remarks
Kangekrullek (Broen) on Sermilik Isle.		Herjulfnaes	C. Farvel	Cape Farewell, 59° 48' N. 43° 53' W.	"Siaten-hük" of Dutch.
Unanarssursuak	"Great heart-shaped."	Hvidsaerk ("white shirt"). 1494, inhabited by the pirate Pinning.			
Kangek	Promontory			Cape Christian, 59° 49' N.	Promontory. So named by Erik Raude. The word means "a place of turning."
WEST COAST OF GREENLAND. From S.					
Kougarsuak	Large cape	Sandhavn. Hvarf			
Ilua	Its interior	Berefjord			The large fiord next west to Hvarf.
Sagdlivik Island	"Where it becomes narrow."	East Bygn.			
Nunarsoak	Illu, "interior."	Spalsund			
Iluamint	Narrow bay (Torssuk, "doorway").	Drangea.			
Torssukatak Strait		Torafjord.			
Hivdlerssunk Bay	Large lowland	Biargtafjord (according to Eggers). Silvudal.			
Igdlukat	Rather large houses.				
Iglorssutsiat		Melrakkanaes			The easternmost settlement.
Pamiagdluk	Pamint, "tail of a seal."				Western point of Frederiksdal.

Native name	Meaning	Place / Church	Identification	Latitude	Remarks
WEST COAST.					
Kvasuk.					
Nahuh Mt.					
Narssak	"Level land"	Herjulsfjord Church.	Fredriksdal, founded 1827 (437 inhabitants).	59° 59' N.	Fredriksdal, 60° N. A Moravian mission.
Ikigait Church (ruins)					
Natssamint Fiord.					
Ujamrasuak	"Big stones"			60° 10' N.	
Korsonk				60° 17' N.	
Ikigait Cape	"Talon"			Ost Proven, 60° 27' N.	
Kitsigaut	Most seaward island.	Holliaoyarfjord			The fjord before you outer Tasermint.
Tasermiut Fiord		Ketilsfjord, Monastery of St. Olaus.		60° 26' N.	St. Olaus', Augustinian Monastery. Church of Aros.
Lake Kangoa		Vatsalal, Petersvig (the chief place).			Large lake.
Napassorsuak Mt. (4000 feet).	"Standing erect" "A hill vomiting fire" (Leao).	Ilmkliurnareya			Extinct volcano.
Nanortalik	"Bear Island" (Nanook, a bear).		Christian IV. Island	60° 8' N.	
Niakornak Point	"Head-like"	Lunduya		60° 14' N.	
Sermersok Mt.				60° 17' N.	
Sermelik	"Having a glacier"	Alptafjord			Supposed by Hans Egede to bo Cape Farewell.
Nunatsiuk	"Nuna," land.			60° 10' N.	
Sermersok Island	Rich in glaciers		Cape Egede		Ice-stream.
Igallorpait	Many houses.		Nicolai Nazo	60° 28' N.	

Name	Meaning			Lat.	Remarks
Unartok Island	"Hot fountains"	Siglufjord Church. (Radlus) .. small islands with hot springs.	Hot springs described by Graah. Temp. 93° 87'. Fahr., another, 102° 87' (Graah, p. 35.)	60° 28' N. Onartok.	Benedictine Convent. Trees 14 feet high.
Agdluitsok	Without breading-holes.	Rafnsfjord	Lichtenau (founded 1772)	60° 31' N.	Moravian Mission.
Knesok Island	Barren		(Sud Proven)	60° 27' N.	
Akuliaruarsenak Mt.	Large promontory with low isthmus.			60° 42' N.	
Umanarsuk Island. Siorllok (Siorllik?)	Narrower part	Stellufjord	Cape Thorvaldsen.	60° 32' N.	
Agdlerusat	"Shape of jaw-bones."				
Kakortok Point	"White"		Julianshaab, the *South*. Colony. Founded in 1776.	60° 42' N. 60° 45' N. (German Exp. p. 220.)	Founded in 1775 by Anders Olsen, whose descendants still reside there.
Kilangak	A puffin		Storefjeld Mountain. Maleuefield .. .. Storöer (great island). Frid. VII. mine.	60° 46' N.	
Pisigsilik	Having a bow		Langöe Island.	60° 45' N.	
Kekertarsuatsink	A lake semicircularly enclosed by a continuous cliff, as the eye by the eyebrow.			60° 47' N.	
Kaqdlumiut	Amilok, narrow "A narrow firth."				
Amitsuarssak Fiord		Hornafjord ..			Large lake.
		Foss, splendid church of St. Nicholas (*Royal demesne*).			Between Igaliko and Agdliutsok (church ruins).

Q

LIST OF NAMES OF PLACES IN GREENLAND—*continued.*

Eskimo.	Meaning.	Norman.	Danish.	Admiralty Chart and Latitude.	Remarks.
WEST COAST.					
Kingigtok	High.				
Arnat	*Arnak,* female.				
Innurulligak	Elf mountain.				
Omenartok.					
Ujararfarfik ..	Place for gathering stones.				
Kekorfassuak.					
Torsukatak.					
Umanarsak.					
Kaersok	Barren rock.				
Umanarssuak.					
Sarilok	Narrower part.				
Upernivik	Place for dwelling in tents during the spring.				
Umanak.					
Omersalik.					
Kangek.					
Igaliko Fiord	*Iga,* pot	Thorvaldsvig .. Einarsfjord ..		:	Arm of the sea left of entrance to Igaliko. Erik Raude's house. Igaliko was the last place occupied by the Normans. Head of Igaliko Fiord.
Kagssiarssuk	Liko an enclosed inlet.	Gardar, town and cathedral. "Gunnuelgord" .. Klining Promontory. Gravavig Promontory.		:	("German Exp." p.229.) Erik Raude's house. {All in the Igaliko Fiord.

Name	Meaning	Other name	Latitude	Remarks
Kangek	..	Renso Island	..	
Akku	..	Lange	..	
Kakortok (ruins of church)	"White" ..	Hvalsöerfjord ..	60° 50' N.	Steatite mines. Here there is a large cavern with a lake in its recesses. Cave 200 feet deep by 32 in width. Kakortok, ruins discovered by Haas Egede in 1723. Visited by Graah in 1829 (p. 38). Church 51 feet long by 25, walls 4 feet thick. Place of assault of the Normans on the Eskimos.
Arpatsivik (Akpeitsivik?)	..	Hvalsöe	..	
Pardliet	Part of a house nearest the entrance.			
Isarut	Wing.			
Kingektok	High.			
Igdlorssuatsiak	Large house.			
Simiutat	Islands sheltering the entrance to a firth, like a stopper (simik).	Skovfjord.		
Niakornak	Niak, head.			
Kingaitsok	Kingak, nose.			
Kaugerdluarsuk	Kangerdluak, bay.			
Igdlokassik	Bad house.	Kaulstndefjord. Tiqulhildtstad (liegd demesne).	60° 53' N.	
Izhutalik	Igha, house.			
Kitlavat Mountain	Indented like a comb.	Redoken Mt. ..	60° 54' N.	..
Narssak	Level land.			
Juktatok	..			

Name	Meaning	Old Norse name	Remarks
WEST COAST.			
Poruk.		Lamboe.	
Irsarut		Lamboesund	Between Irsarut and Akkia.
		Hafgrimsfjord	Head of Igaliko Fiord.
		Eskarsfjord	
		Church.	
Tunnudliarfik	A place for going to the interior land.	Eiriksfjord.	On the left hand as you enter Tunnudliarfik (ruins).
Tugtutok	Tuktu, reindeer. "Rich in reindeer."	Eiriksöö.	The broad fiord leading N.W. from Tunnudliarfik.
		Dyrnaes Church	Further inside. Ruins.
		Midfjord	Still further inside.
		Solfjall Church	On S. side of fiord of Tunnudliarfik.
		Leide Church	Episcopal steward's residence.
		Burfjall	
Karmangnak	Small wall of a house.	Brattalid	
Akunelt	"Lying between."	Fossa Sund.	
Ukivisoknak	Old wintering place.		
Kangatsiak	Promontory.		
Iglu	House.		
Sermilik	Having a glacier.		

Name	Meaning	Norse name		Note	Remarks
Ikersunk	..		..		{ Two inlets south of Tuktutok.
					{ Two fiords next west to Ikersunk.
Kagssimiut	A house for public assemblies.		..		
Kangerdluluk	Kangerdluk, bay.	Indrevig .. Ydrevig .. Brediffjord .. Eyrarfjord .. Borgofjord ..	..		
Imartinek	Large inlet ..				
Sermitsialik	Having somewhat of a glacier.	Lodmundarfjord	..		Last place in the East Bygd.
Kingitok	Kingitok, high.	Isafjord ..			
Kakaligatsiak	A rock of peculiar shape.				
Kekertatsiak	Island.				
Sarkak	"Sunny side."				
Umanak					
Nunatsuit (highest summit)	Great lands ..		..		C. Desolation of Davis, according to Graah. 60° 43' N, Graah.
Kitdlavat	Indented like a comb.		..		
Agdlerussat	Like a jaw-bone.				
Tasiussak	Like a lake.				
Amitsuarsuk	Amitdok, narrow inlet.		..		
Torsukatak Strait	Torsuak, doorway.		..		
Itinilliatisiak	Narrow isthmus ..				
Kitsigsunguit Kitdlit	Outermost of outer islands.			Iosuns mine ..	Aurora's Harbour, 60° 49' N. 60° 55' N.
Uuanarsuk			..		
Kitsigsunguit kangigalit	Innermost of small outer islands.		..	Thorstein Islands	Thorstein Islands, 60° 48' N. Storö Isle.
Agssunguit Strait	Small fingers.				
Kepisako					

LIST OF NAMES OF PLACES IN GREENLAND—*continued.*

Eskimo.	Meaning.	Norman.	Danish.	Admiralty Chart and Latitude.	Remarks.
WEST COAST.					
Ilimansak	Like small peg (*ilimak*) on harpoon handle.		Narksakfield	60° 59' N.	
Sauerât Island	Crosspiece				
Kakortok Point	White			61° 0' N.	
Umanak	With an edge.				
Kinalik					
Kuiartorfik	Tattoo marks.				
Tavdlorutit					
Kajartalik		West Byan.		61° 10' N.	Begins in 69° 30' N. of Iceblink of Fredriikshaab.
Arsuk	Spring camping ground.	Steinmacs (church)			
Upernivik Point				Assut, 61° 16' N. 61° 11' N.	
Kunak Mountain	*Iwigloi,* duck.			61° 15' N.	Cryolite mines.
Iwiglot, or .. *Iwikat* (Arsot Fiord) (Arksut?)	*Iwik,* sea-horse		Smallo-sand	61° 34' N.	Ice stream.
Sermeliansuk	Magic spell.			61° 32' N.	
Serrat					
Umanak				Cape Comfort, 61° 49' N.	
Sermiligatsiak	Having a glacier	Heemelmchs Mountain to the north of the West Bygd; beyond which no man may sail without peril of his life, on account of the numerous whirl-		Tiksalik Point.	

Name	Meaning	Founded / Mission	Position	Notes
Kangârssuk	Promontory.			
Sermilik	Having a glacier.		61° 57' N.	Ice stream.
Narsalik	Having level low-land.			
Umanak.				pools in that sea. —Ivar Bardsen. But in earlier times the Normans had habitually gone far north of this.
Kuanersôk	Rich in Angelica Archangelica.			
Pâmiut	Inhabitants at the entrance of a firth.	Fredrikshaab, founded 1712.	Fredrikshaab, 62° N.	Here the glacier projects a mile out to sea.
Nerutusok	Wide, capacious.		Sinaskanner Island.	
Avigait				
Tekkisok				
Siorak	Sand		Karakotfjord.	
Ikatok	Shallow water		Tallart bank.	Kär Sund, formerly a passage through to the E. side (Crantz, vol. i. p. 6).
Agdlumersat.			Kikertar Isle.	
Sarfak	Current..		Biorno Sound.	
Kangâtsiak				
Fullarlalik Isles	Hooded seal.		62° 32' N.	
Nat-erssuak				
Kangigdlit	Those most inland	Fiskernæs	Fiskernæs, 63° 4' N. (Grnah), 63° 7' N. (Inglefield).	Fischer Fiord Crantz.
Kikertarssuatsak Island	Outermost, seaward and middlemost.	Lichtenfels. Moravian Mission, founded 1758.	Lichtenfels, 63° 5' N.	
Kitdlit-akungnât				
Kangârssuk	Cairn		Kangârsuk	
Iluugsuk (Crantz, i. p. 7)			Ikersar Fjord.	Kollingeit Islands (Crantz, vol. i. p. 7).
Kangdluarssugssuuk	Firth			

LIST OF NAMES OF PLACES IN GREENLAND—continued.

Meaning.	Norman.	Danish.	Admiralty Chart and Latitude.	Remarks.
Sermik, ice		..	Trebröire Havn.	Baxe Bay (Crantz, vol. i. p. 8).
Brothers.				
Barren			Merkorit Fjord.	(Stag's Horn), one of the highest mountains in Greenland (Crantz).
Having a cover ..	Norman church (ruins).			
Cape.				
Narrow part of} sound.	..		.. ..	(Kellingarsoak, (Kariak
			Gilbert Sound of Davis. Störo.	Bull's River.
Large salmon ..	..	Godthaab	..	Norman runes. Ice stream.
Northernmost ..	..	Vildmansnces (Hans Egede).	Godhuah, 64° 10' N. 51° 45' W.	Godhaab (founded 1721 by Hans Egede).
Those nearest the point.	..		New Herrnhuth.	New Herrnhuth. Moravian Mission, founded 1761.
Place for shooting with bows.	..		Pisuklit Isles.	
Red.				
Large stone	Norman church (ruins).	..	Ujarak Fiord.	
Fish.				

Native name		Meaning		Modern name	Coordinates / identification	Remarks
Niakungunak. Atangunik	.. ..	Long point of land.	..			
Nappersok	..	A pole.			Nappasok Fiord. 65° 25' N.	Nappasok.
Umanak.	..	Muddy water from glaciers.				
Isortok	..					
Manitsok	..	Uneven, rugged .	...	Ny Sukkertoppen	...	Several other places of that name.
Unanarssuak. Sermersok. Agnamiut. Kangerdlugisuatsiak. Kangâmiut.	.. ..	"Inlet."				
Kangerdlugssuak	..	"The high island."			65° 25' N.	"Wyde Fiord" of the Whalers.
Umanarssuak	..	Cape, headland.	Lysefjord ..	Sukkertoppen, founded in 1755.	Sukkertoppen, 65° 20' N. Coquin Sound, 65° 38' N. Kikertak Island. ItiblikfJord.	
Kaugôk	..	"Loom's tail " ..				
Ag-pamiut			...		Cape Anna, 66° 24' N.	
Itiollek (Hivdlingnak, Rink). Ikertok	..	Wide firth to cross	...			
Narsamiut. Amerdlok Sisimiut		Narrow firth	..	Holsteinborg, founded in 1759. Sir John Ross says the native name is *Triraiak Pudlit* ("foxes' holes"). It should be *Teriuniak* (a fox), and *Tudilet* (a fox-trap).	Holsteinborg, 66° 58' N. 53° 54' W. (Sir J. Ross).	Christians' Fjord. Ramel Fjord. Mt. Cunningham.

LIST OF NAMES OF PLACES IN GREENLAND—continued.

Eskimo.	Meaning.	German.	Danish.	Admiralty Chart and Latitude.	Remarks.
WEST COAST.					
Isortok Island .. :	Muddy water			Cape Sophia, 67° N. *Division between North and South Greenland.* Cunningham Fjord. Isertok Island.	
Nagsutok :	*Nagsuk*, horn			Cape Amalia. Simiutak Island.	
Akto (Atto, Rink) .. :		:		Neksotuk Fjord. Ekallug Island. Aito. Riftkol.	Riffkull (Crantz, vol. i. p. 14).
Umanak. Iginiarfik .. :	Place for shooting	Karlsbudas.		Iginiarfik.	
Aulutsivik .. :	Place for being moved (by the sea).	:		Aulcitsivik Fjord. 68° 38' N.	
Satut .. : Kangatsiak	Flat.	:	Vester Island.		
Kokertarsitsiak : Umivik .. : Manermiut .. : Ausinit .. :	Landing-place. Moss. "Spider"	:	Egedesminde, founded in 1750.	Kangeitsiak. Narsarsuk. Kepingosok. Kangarsutsiak.	
Mauitsok Island :	Rugged.		Hunde Island :	Egedesminde. 68° 46' N. (?) (68° 43' N. *Graah.*) Hunde Island, 68° 50' N.	
Kitsigsorsuit .. :	Large outer islands		Kron Prins Isles .. :	Whale Fish Islands, 68° 57' N.	

Akunak Islands	Middlemost	::	..	Akunak Islands.
Itidlermiut.				
Kernatissortik.				
Upernivik Isle.				
Tosak Isle.				
Akulick Isle ..	Middlemost.			
Sarpiursak ..	Like a whale's tail	::	Gronne Islands	Gronne Islands. Sarpiursak. (S.E. Bay).
Orpik<oit ..	Orpik, bushes { properly the birch. *Orpik*, a bush			
Kasigianguit ..	*Kasiguak*, Phoca barbata.	::	Christianslaub, founded in 1734.	Christianshaab, 68° 49′ N.
Kekertarsuaduk.				
Ikamiut.				
Kasigianguit.				
Itimanak		::	Clausbavn, founded in 1752.	Clausbavn, 68° 7′ N.
Ilulissat (Iluligsat, Rink)	"Place of icebergs"			*Ice stream.*
Tessuisak.	Peak of land emerging from a glacier.	::	Jacobshavn, founded in 1741, by Jacob Severin.	Jacobshavn, 69° 12′ N. Rode Bay.
Nunatak	Innermost.			
Illorilek	Narrow entrance ..			Pakitsok.
Pakitsok	.. ..	::		Ikaresak Sound.
Ikarsak Island	.. ..	::		Prinzdens Island.
				Arva Prince's Island (Brown).
Agpat	Looms	::	Rittenbenk, founded in 1755.　69° 48′ N.	Klokkerbuk. Rittenbenk. 69° 41′ N. Tossukatck, *ice stream.*
Tossukatak (ice stream)				
Tasersuak-Indso ..	Great lake.	::		Rifdliarsuk.
Itivdliarsok ..				
Majorkarssuatsiak ..	Road for going up a mountain.			

LIST OF NAMES OF PLACES IN GREENLAND—*continued.*

Eskimo.	Meaning.	Norman.	Danish.	Admiralty Chart and Latitude.	Remarks.
WEST COAST.					
Kekertak				Kekertak.	
Sarkak	Sunny side			Saklak.	
Atanikerdluk	Joined to main land			Atanekerdluk, 70° 2' 30" N.	
Tatarak (Iecalland)..	Kittiwake.				
Ekaluk (creek) ..	*Ekaluk*, salmon.				
Marrak	Clay.				
Kugssuak	River.				
Taserssuak (lake) ..	Lake.				
Manik	*Manuek*, moss.				
Atane	Beneath.				
Ninkornarsuk	Head.				
Itivdlek					
Nugssuak	Great point.	Kysunaes	Noogsoak, founded 1758 (Crantz).	Noursoak peninsula.	
DISCO.				Waygat.	
Kekertarssuak ..		Bjarney	Disco	Disco.	Western coast of Disco, called by the whalers "*Bunky Land*."—(R. B.)
Huitek } Kaersok }	Peninsula .. .. Bare rock ..		Godhavn	Leively, 69° 15' N. (69° 14' N. *Graah*). (69° 4' N. *Inglefield*.)	
Sinigfik	Place for sleeping	"Nordrsetur," discovered by a party of Norman clergymen from Green-			

Name	Meaning	Notes	Locality
Ujararsunk	Ujarah, stone.	land, in about 1256, is supposed to have been Disco. There is mention of places called "Greipar," and "Kroksfiardarheidi" in these parts, as constantly visited by the Normans.— R. G. S. J., vol. viii., p. 126.	
Assuk			
Kingua	Interior part of firth.		
Igænek	Cooking.		
Sarfarssuit	Strong currents	Skandsen. Klokker huk	Flakker Point, 61° 32' N. 63° 51' N. (Akajarua ?)
Uivfait	Ferns.		
Kakak Island	Kakkak, hill		Igler Point. North Fiord. Mellem Fiord.
Nulok Island	Buttocks		
Kekertarssuatsiak			Kakkak Island. Nulluk Island. Hare Island, N. Pt., 70° 29' N.
Aumarutigssat	Coals.		
Unvit			
Sermesok	Sernik-sunk, "Great ice" (Glacier).		Omenak Fiord.
Niakornak			Niakornak, 70° 47' N.
Tunugssuak	Large inland valley.		"Four Islands Point" of whalers.

LIST OF NAMES OF PLACES IN GREENLAND—continued.

Eskimo.	Meaning.	Norman.	Danish.	Admiralty Chart and Latitude.	Remarks.
Iborput Point .. .. ::	A prop.	Kindelfjord ::	Omenak ::	Omenak, 70° 40′ N. (70° 41′ N. Greah).	Danish settlement.
Umanak .. .. ::				Ikaresak.	
Iketrisuk .. ::	Bay surrounded by cliffs.			Kariak Glacier. 70° 26′ N.	Ice stream.
Karajak Glacier (Karak) ..					
Sernuelik Glacier ::	"Sermik," ice.			Sernuelik Glacier. 70° 41′ N.	Ice stream.
Innerit .. ::	Fire	:: :: ::		Innerit. 70° 56′ N.	Ice stream.
Uvkusigssak ::	Soapstone			Okosiksak.	
Akpat .. ::	Loous			Akpet.	
Satut .. ::	Flat			Saitut.	
Sagullerssuak ::	Foremost.				
Upernivik ::	Tent place during spring.			Upernivik.	
Kangerdlursuak Glacier	Large firth .. ::		Ubekionlt Island ..	Ubekionlt Island. Kangerdlursoak	::
Kekertarssuak ..	Kekertak, island .. kekertarssoak; great island.		::	Glacier. 71° 25′ N. Kikertarsoak.	Ice stream.
Kakak ::	Kakak, hill ..	Œdannoes :: ::		Kakkak. Svarte-luk, 71° 38′ N. (Greah) (71° 25′ N. Inglefield). Skali. Kingitak.	
Kingarfak .. ::	Sharp side of shintone.		::	Kikerstar Island.	
Kekertak Island .. ::			Proven :: ::	Proven, 72° 21′ N. Greah. (72° 32′ N. Inglefield.)	Danish settlement.

Ikerasak ..	..	..	..	Ikarssak.	
Nutarmiut	..	..	..	Nutarmiut.	
Akkaliarrusek	..	..	..	Akkaliarrusek.	
Opersoarsoit	..	..	..	Opersoarsoit.	
Nekitarssok	New ..	..	..	Nakitarssok.	
Kasorsoak,	Promontory ..	..	..	Hope Sanderson, 3300 ft.	Danish settlement.
	Pressing, kneading				
Upernivik..	"Spring-place." "Upernuk," spring.	..	Upernivik	Upernivik, 72° 48' N. (Sun 79 days below horizon).	Upernivik and surrounding islets are the "Women Islands" of Baffin.
Ikak, Tharsok. Tessiussak. Kingigtok Island (Kingiktorssuak ?).	Kingít-pok, it is high.	A very remarkable runic stone was found on the island of Kingiktorsuak in 1824.— R. G. S. J., vol. viii., p. 127.	Kingitok	Kingitok, 72° 55' N.	Danish settlement.
Augpadlartok ..	Red ..	..	..	Aukpadlartok. Brown Islands. Berry Islands, 73° 29' N. Wedge Island. Cone Island. Resolute Rock.	
Country uninhabited from Kingigtok to Cape York, 250 miles in a straight line, or 300 miles round the shore.				MELVILLE BAY. Cape Shackleton, 1400 ft. Horse's Head. Baffin Islands. Duck Islands. Wilcox Head. Devil's Thumb. Allison Bay.	

LIST OF NAMES OF PLACES IN GREENLAND—*continued.*

Eskimo.	Meaning.	Norman.	Danish.	Admiralty Chart and Latitude.	Remarks.
COUNTRY OF THE ARCTIC HIGHLANDERS.[1]					
Kugssuak Glacier ..	Great river ..	The party of Norman clergymen from Gardar, who made a voyage in 1266, are calculated to have reached 75° 46' N. in the neighbourhood of Cape York. They called a cape "Sniofell".— R. G. S. J., vol. viii., p. 126.	..	Cape Seddon. Cape Lewis. Melville Monument. Sabine Islands. Cape Walker. Leven and Balgonio Islands. Cape Murdoch. Cape Melville. Bushman Island.	
Ignigsok Immangen (E.Y.)	Rich in walls of clay	..	..	Cape York, 75° 52' N. (73° 56' N., Kane).	

Native name	Meaning	English name / Position	Remarks
Imatik (K.)	Full.		
Ahpa (E.Y.)			
Kokerlut (H.)	Islands.		
Savik (iron mine)	Iron		
Jenna (E.Y.)		Parker Snow Point, 76° 4' N. Cape Dudley Diggos, 76° 20' N. Saunders Island.	Plenty of hares.
Agpa (E.Y.)	"Loom"		
Sernliksuak (E.Y.)	"Great ice."		
Kaersorsuak (E.Y.)	Great barren rock.		
Kauuti (E.Y.)			
Kaugarsuk (E.Y.)			
Kekerlassuak (E.Y.) Agpa	"Great Loom Island."	Cape Atholl. Wolstenholme Island. Dalrymple Rock, 76° 30' N.	Valley torrent during summer.
Jenna (E.Y.)			
Narssarsuak (E.Y.)	Large level land..		
Agpa (E.Y.)	"Loom."	Wolstenholme Sound, 76° 31' N. North Star Bay, 76° 33' N.	Deer in plenty. Plenty of lime-stone. Conical Island.
Appak (K.)			
Umanak (E.Y.)			
Igluimanak (E.Y.)	"Woman House"		
Izuoi (E.Y.)			
Jenna (E.Y.)			
Tasiussak (K.)	Like a pond.		
Kaugarsuk (E.Y.)		Cape Parry.	
Natsilik (E.Y.)	Having seals.	Whale Sound, 77° 5' N.	York did not know whether this Sound continued far up or not.
Itibla (E.Y.)			
Iteplik (H.) Itivdlek?	Inner side.		
Kanga (E.Y.)			

R

Native name	Meaning	European name	Remarks
Kangerlugssuak (E.Y.)	Large bay.		
Kavra (E.Y.)	South side.		
Sittoni (E.Y.)			
Ionagnuem (E.Y.)	Inuak, precipice.		
Kanga (E.Y.)			
Agpagssuak (E.Y.)	"Loom," great	Herbert Island.	High up Murchison Sound.
Kujata (E.Y.)	South of it	Hakluyt Island. / Northumberland Island.	Abounds in hares.
Neko (E.Y.) / Natsilik (K.)	Meet / Having seals	Murchison Sound, 77° 7' 8" N.	
Ujarsuk Suk-suk (K.)	Ujaruk, stone		
Karsuit (K.)		Cape Robertson.	
Iglunakstuk (E.Y.)		Cape Saumarez, 77° 30' N.	(L.)
Arwagtuarwi (E.Y.)			
Pitorak (E.Y.) / Pitoralik (K.)	A sudden strong wind. Where sudden strong winds are blowing.	SMITH SOUND.	
Siomlik (E.Y.)	Having sand.		
Uglaksun (E.Y.) / Kudluktunrseuk (E.Y.)			A small fresh water lake here, containing fish.
Ekaluk (E.Y.)	Salmon		Many wolves about this island.
Pikieriu (E.Y.)			

Cape Alexander, 78° 12' N. (78° 9' N. Kane.) Cape Ohlsen, 78° 17' N. Port Foulke, 78° 17' 39" N. Cape Isabella, 78° 22' 15" N. Littleton Island, 78° 22' 5" N.	(*Haye's Winter Quarters*, 1860.)
Rensselaer Harbour, 78° 38' N. Cape Frazer (79° 42' N. *Kane*).	(*Kane's Winter Quarters*, 1853–55).
	A traditional island in the far north.

II

ON THE DESCENT OF THE ESKIMO.

By Henry Rink, Director of the Danish Colonies in Greenland.[1]

The author, who has travelled and resided in Greenland for twenty years, and has studied the native traditions, of which he has preserved a collection, considers the Eskimo as deserving particular attention in regard to the question how America has been originally peopled. He desires to draw the attention of ethnologists to the necessity of explaining, by means of the mysterious early history of the Eskimo, the apparently abrupt step by which these people have been changed from probably inland or river-side inhabitants into a decidedly littoral people, depending entirely on the products of the Arctic Sea; and he arrives at the conclusion that, although the question must still remain doubtful, and dependent chiefly on further investigations into the traditions of the natives occupying adjacent countries, yet, as far as can now be judged, the Eskimo appear to have been the last wave of an aboriginal American race, which has spread over the continent from more genial regions, following principally the rivers and water-courses, and continually yielding to the pressure of the tribes behind them, until at last they have peopled the sea-coast.

In the higher latitudes, the contrast between sea and land, as affording the means of subsistence, would be sufficient to produce a corresponding abrupt change in the habits of the people, while further to the south the change would be more gradual. The water-courses which may have led the original inland Eskimo down to the sea-coast might probably have been the rivers draining the country between the Mackenzie and the Athna rivers (? Athabasca).

The same country also seems to afford the most probable means of explaining the uniformity observable in the development of Eskimo civilisation, which to some extent is still maintained amongst

[1] An article in the 'Mémoires de la Société Royale des Antiquaires du Nord.' (From the 'Journal of the Anthropological Institute.' April 1872.)

them upon the rivers and lakes in that part of America. This development must have been promoted by the necessity of co-operating for mutual defence against the inland people; but as soon as a certain stage of development was attained, and the tribes spread over the Arctic coasts towards Asia on the one side and Greenland on the other, the further improvement of the race appears to have ceased, or to have been considerably checked.

The author draws a comparison between the Eskimo and the nations adjoining them, both in Asia and America, in regard to their arts of subsistence, language, social laws, customs, traditions, and other branches of culture, particularly dwelling on their traditions, of which he has collected a great number from all the inhabited places on the east side of Davis Straits, together with some from East Greenland and Labrador. He shows that an astonishing resemblance exists between the stories received from the most distant places, as, for instance, between those of Cape Farewell and Labrador, the inhabitants of which appear to have had no intercourse with each other for upwards of a thousand years. As the distance from Cape Farewell to Labrador, by the ordinary channels of Eskimo communication, is as far as from either of those two places to the most western limit of the Eskimo region, it may be assumed that a certain stock of traditions is more or less common to all the tribes of Eskimo. The author's studies have led him to the following conclusions: 1. That the principal stock of traditions were not invented from time to time, but originated during the same stage of their migrations, in which the nation developed itself in other branches of culture; viz., the period during which they made the great step from an inland to a coast people. The traditions invented subsequent to this are more or less composed of elements taken from the older stories, and have only had a more or less temporary existence, passing into oblivion during the lapse of one or two centuries. 2. That the real historical events upon which some of the principal of the oldest tales are founded, consisted of wars conducted against the same hostile nations, or of journeys to the same distant countries; and that the original tales were subsequently localised, the present narrators pretending that the events took place each in the country in which they now reside—as, for instance, in Greenland, or even in special districts of it. By this means it has come to pass that the men and animals of the original tales, which are wanting in the localities in which the several tribes have now settled, have been converted into supernatural beings, many of which are now supposed to be occupying the unknown regions in the interior of Greenland.

In accordance with these views, the author explains some of the most common traditions from Greenland as simply mythical narrations of events occurring in the far north-west corner of America, thereby pointing to the great probability of that district having been the original home of the nation, in which they first assumed the peculiarities of their present culture. The Greenlander's tales about " inland people " are compared with what is known about the present intercourse of the Eskimo with the interior of that part of America, such as instances of relationship between the people of the coast and the interior, sudden and murderous attacks of the latter, and a very remarkable story about an expedition to the interior for the purpose of getting copper knives from the inland people. Lastly, there are some tales about the country beyond the sea called Akilinek, and about the training of wild animals for sledge expeditions to this country, in order to recover a woman carried off by some inhabitants of that country. When we consider the existing intercourse between the inhabitants on both sides of Behring Straits, we find many circumstances to justify the conclusion that those traditions of the Greenland Eskimo refer to the origin of the Eskimo sledge-dog from the training of the Arctic wolf, to the first journeys upon the frozen sea, and to intercourse between the aboriginal Eskimo and the Asiatic coast.

III

THE WESTERN ESKIMO.

Observations on the Western Eskimo, and the Country they inhabit; from Notes taken during two years at Point Barrow. By Mr. John Simpson, Surgeon, R.N., Her Majesty's Discovery Ship, *Plover.*[1]

The term Western Eskimo is usually understood to apply to all the people of that race who are found to the west of the Mackenzie River, but as they form two distinct communities, whose nearest respective settlements are separated by an interval of three hundred miles of coast, it is proper to state that the term is at present restricted to the more western branch. The tract of country exclusively inhabited by them is that small portion of the north-western extreme of the American continent included by a line extended between the mouth of the Colville River and the deepest angle of Norton Sound, and the coast-line from the latter through Behring Straits and the Arctic Sea back to the Colville. The seaboard for a little way to the south of Norton Sound is also occupied by a few scattered families of the same race. As these people divide themselves into numerous sections, named after the portions of land they inhabit or the rivers flowing through them, it will be convenient, before speaking more particularly of themselves, to give some account of the country as described by them. The information is principally derived from the people of Point Barrow, some of whom have travelled and lived for a time in different localities, and from strangers who came to visit them during the time of the *Plover's* stay at that place.

By Captain Beechey's survey, the south and western part of this district will be seen to be mountainous and deeply indented by arms of the sea, but the northern and more inland portions have been examined to only a short distance from the coast. The natives of Point Barrow describe the latter as uniformly low, and full of small lakes or pools of fresh water to a distance of about fifty miles

---

[1] From 'Further Papers relative to the Recent Arctic Expeditions in search of Sir John Franklin.' Parliamentary Reports, 1855.

from the north shore, where the surface becomes undulating and hilly, and, farther south, mountainous. The level part is a peat-like soil covered with moss and tufty grass, interspersed with brushwood, perfectly free from rocks or stones, and only a little gravel is seen occasionally in the beds of rivers. The bones of the fossil elephant and other animals are found in many localities, and the tusks of the former are used for some purposes. Small pieces of amber are also frequently found in the pools inland, or floating on the sea, to which they have been carried in the summer by the floods. The whole is intersected in various directions by rivers, which are traversed by boats in the summer and by sledges in the winter. Many of the streams seen from the coast become united, or have a common origin in some pool in the interior, and sometimes offer a short channel from bay to bay, deep enough for boats, which thus avoid a more circuitous and inconvenient passage round the coast.

The largest and best known rivers are four, all of which take their rise far to the south-east in a mountainous country, inhabited by Indians. The most northerly of these is the Kang'-e-a-nok, which flows some distance westward, then turns northward, receiving on its right bank two tributaries, called the A'-nak-tok and Kil'-lek. At a distance of probably one hundred miles from the coast it divides into two streams, the eastern of which follows a nearly north course to the Arctic Sea, one hundred and forty miles east of Point Barrow, where it has been identified with the Colville. It bears the native name of Nig'-a-lek Kŏk, or Goose River, and is said to receive a large tributary at thirty miles from its mouth, called the It'-ka-ling Kŏk, or Indian River, coming in from the mountains in the east. The other division flows through the level country nearly due west to fall into Wainwright Inlet, ninety miles S.W. of Point Barrow, when it is named Tu-tu-á-ling, but is more generally known as Kŏk or Kŏng, "the River." The next is called the Nu-na-tak', also a large river, whose source is very close to that of the Colville ; but instead of turning, like the latter, northward, it pursues a westerly course through the heart of the country ; then, bending to the south and a little east, falls into Ilotham Inlet, near its opening into Kotzebue Sound. This certainly, in the estimation of the Point Barrow people, is the most important river in their country, and gives its name to by far the larger portion of the inhabitants of the interior. At one point of its course it approaches so near a bend of the Colville that boats can be trans-ported in less than two days from one river to the other. The Kó-wak is the next in order as well as in size and importance, chiefly on account of a few mineral substances procured in its neighbourhood, and held in esteem by the natives of the coast. It also flows westward, and then bends southward to join Hotham Inlet near its eastern end. The fourth is the Sí-la-wik, which, having a more southerly origin, follows a more direct westerly course, and empties itself into a large lake, communicating with the eastern extreme of the same inlet near the mouth of the Kó-wak. All these rivers have been identified by different officers from the *Plover* having visited their embouchures, and those falling into Hotham Inlet were found bordered with large pine trees. The natives add, that trees also grow on the banks of the rivers in some parts of the interior. The other rivers along the north and north-west coast are small and hardly known, except to persons who have visited them : and the Buckland and others to the southward are but little spoken of by the people generally, although aware of their existence.

The largest settlements are at Point Barrow, Cape Smyth, Point Hope, and Cape Prince of Wales, which are never altogether deserted in the summer ; but besides these, there are numerous points along the coast, as at Wainwright Inlet, Icy Cape, the shores of Kotzebue Sound, Port Clarence, and Norton Sound, where there are smaller settlements or single huts, occupied in the winter but generally abandoned in the summer.

The inhabitants state, that the sea affords them several varieties of whale, only one of which is usually pursued, the narwhal (occasionally), the walrus, four different sorts of seal, the polar bear, and some small fish ; the inlets and rivers yield them the salmon, the herring, and the smelt, besides other kinds of large and small fish ; and on the land, besides abundance of berries and a few edible roots, are obtained the reindeer, the inna (an animal which nearly answers to the description of the argali or Siberian sheep), the hare, the brown or black bear, a few wolverines and martens, the wolf, the lynx, blue and black foxes, the beaver, musk-rats and lemmings. In summer, birds are very numerous, particularly geese in the interior and ducks on the coast. The ptarmigan and raven remain throughout the winter, and the latter is the only living thing we know to be rejected as food. Black-lead, and several varieties of stones for making whet-stones, arrow-heads, and labrets, and for striking fire, are also enumerated as the produce of the land and articles of barter. The articles in common use, for which they are indebted to strangers, are kettles, knives, tobacco, beads, and tin for making pipes, almost all of which come from Asia. English knives and beads are also in use, and within these few years, at Point Barrow, the Hudson's Bay musket and ammunition. The

skin of the wolverine is held in high esteem, and is, like the
English goods, procured from the Indians, occasionally directly, but
most commonly through their more eastern brethren at Barter
Point. The latter also supply narwhal-skins, large lamps or oil-
burners, made of stone, which form part of the furniture of every
hut.

The great trading places are King-ing, at Cape Prince of Wales,
Se-en'-a-ling, at the mouth of the Nu-na-tak, Nig'-a-lek, at the
mouth of the Colville, within their own country; and Nu-wú-ak,
at Point Barter, to the eastward, between all of which there is a
yearly communication. It might be expected that the Russian ports
near Norton Sound would supply the Russian goods, but such is not
the case, as they are all, or nearly all, brought from the Kokh'-lit
Nuna, as they call Asia. They say four or five Asiatic boats cross
the Straits after midsummer, proceeding from East Cape to the
Diomede Islands, and thence to Cape Prince of Wales, where trade
is carried on with people belonging to the neighbourhood of Norton
Sound, Port Clarence, &c. The boats then proceed along the shore
of Kotzebue Sound until the high land, near Cape Krusenstern
comes into view, when they steer by it for Hotham Inlet, and
encamp at Se-sú-a-ling. At this place, towards the latter end of
July, people from all the coast and rivers to a great distance meet,
and an extensive barter takes place among the Esquimaux themselves, as well as with the Asiatics, amid feasting, dancing, and
other enjoyments. A large proportion of the goods falls into the
hands of the people living on the Nu-na-tak, who carry it into the
interior, and either transfer it to others, or descend the Colville
with it themselves the following year, to meet their friends from
Point Barrow. At the Colville the same scene of barter and
amusement takes place in the latter part of July, and early in
August the goods are carried to Point Barter by the Point Barrow
traders, to be exchanged for the English and other produce of the
east. The Nu-na-tung'-meun, or Nu-na-tak people, thus become
the carriers of the Russian kettles, knives, &c., to be found along
the north coast, and being known only by name to the inhabitants
east of the Colville as the people from whom these articles are procured, it is easy to perceive how Sir J. Franklin and Mr. Simpson
were led to conjecture that a Russian port existed upon that river,
and that the agents residing there were called Nu-na-tang'-meun.
The word Nu-na-tak appears to signify "inland," from its being
commonly applied to persons coming from any part of the interior;
but they do not use any corresponding word to comprehend the
different tribes on the coast.

The number of inhabitants within the first-named boundaries does not, from all we can learn, exceed 2500 souls, and is probably little more than 2000, all of whom have the same characteristics of form, feature, language, and dress, and follow, with little variation, according to the locality, whether on the coast or in the interior, the same habits and pursuits. The remarks which follow, therefore, though more particularly referring to the people of Point Barrow, will be equally applicable to them all.

Point Barrow is the northern extreme of this part of the American continent, consisting of a low spit of sand and gravel projecting to the north-east. Its length is about four miles, and it is little more than a quarter of a mile in average breadth, but expands considerably at the extremity, where it rises to about sixteen feet in height, and sends out to the E.S.E. a low narrow ridge of gravel to a distance of more than two miles, succeeded in the same direction by a row of sandy islets, enclosing a shallow bay of considerable extent. The assemblage of winter huts is placed on the expanded and more elevated extremity, where there is a thin layer of grassy turf. It is called Nu-wuk, or Noo-wook, which signifies emphatically "The Point." No doubt the settlement owes its existence to the proximity of the deep sea, in which the whale can be successfully pursued in the summer and autumn, and to the great extent of shallow waters around, where the seal may be taken at any season of the year. The number of inhabited huts in the winter of 1852-3 was fifty-four, reduced to forty-eight in the succeeding year in consequence of the scarcity of oil to supply so many fires, besides a few others which do not seem to have been tenanted for several years, and two dance-houses. The total population at the end of 1853 was 309, of whom 166 were males and 143 females. The older people say their numbers are much diminished of late years, a statement to the truth of which the remains of a third dance-house and the number of unoccupied huts bear silent testimony. The latter are in some degree taken care of as if to preserve the right of ownership, and to prevent their being pulled down. Further, a disease, which from description seems to have been influenza, is said to have carried off no less than forty people in the commencement of the winter of 1851-2. In 1852-3 the births we heard of were four or five, and the deaths about ten ; and within the last twelve-month, when our information was more accurate, we noted only four births, but no fewer than twenty-seven deaths, most of which occurred from famine, reducing the population at the present time to 286. The settlement at Cape Smyth, about ten miles distant, consisting of forty huts, and having about three-fourths the inhabitants, has been reduced in a more than proportionate degree, having lost forty people since July 1853. Some of these had fled in the depth of winter from their own cold hearths to seek food and warmth at Nu-wûk, where, finding no relief, they perished miserably on the snow. These people are by no means the dwarfish race they were formerly supposed to be. In stature they are not inferior to many other races, and are robust, muscular, and active, inclining rather to spareness than corpulence. The tallest individual was found to be 5 feet 10½ inches, and the shortest 5 feet 1 inch. The heaviest man weighed 195 lbs., and the lightest 125 lbs. The individuals weighed and measured were taken indiscriminately as they visited the ship, and were all supposed to have attained their full stature. Their chief muscular strength is in the back, which is best displayed in their games of wrestling. The shoulders are square, or rather raised, making the neck appear shorter than it really is, and the chest is deep; but in strength of arm they cannot compete with our sailors. The hand is small, short, broad, and rather thick, and the thumb appears short, giving an air of clumsiness in handling anything; and the power of grasping is not great. The lower limbs are in good proportion to the body, and the feet, like the hands, are short and broad, with a high instep. Considering their frequent occupations as hunters they do not excel in speed, nor in jumping over a height or a level space, but they display great agility in leaping to kick with both feet together an object hanging as high as the chin, or even above the head. In walking, their tread is firm and elastic, the step short and quick; and the toes being turned outwards and the knee at each advance inclining in the same direction, give a certain peculiarity to their gait difficult to describe.

The hair is sooty black, without gloss, and coarse, cut in an even line across the forehead, but allowed to grow long at the back of the head and about the ears, whilst the crown is cropped close or shaven. The colour of the skin is a light yellowish brown, but variable in shade, and in a few instances was observed to be very dark. In the young, the complexion is comparatively fair, presenting a remarkably healthy sunburnt appearance, through which the rosy hue of the cheeks is visible; before middle life, however, this, from exposure, gives place to a weather-beaten appearance, so that it is difficult to guess their ages.

The face is flat, broad, rounded, and commonly plump, the cheek-bones high, the forehead low, but broad across the eyebrows, and narrowing upwards; the whole head becomes somewhat pointed towards the crown. The nose is short and flat, giving an appearance of considerable space between the eyes. The eyes are brown, of different shades, usually dark, seldom if ever altogether black, and generally have a soft expression; some have a peculiar glitter, which we call gipsy-like. They slope slightly upwards from the nose, and have a fold of skin stretching across the inner angle to the upper eyelid, most perceptible in childhood, which gives to some individuals a cast of countenance almost perfectly Chinese. The eyelids seem tumid, opening to only a moderate extent, and the slightly arched eyebrows scarcely project beyond them. The ears are by no means large, but frequently stand out sideways. The mouth is prominent and large, and the lips, especially the lower one, rather thick and protruding. The jaw-bones are strong, supporting remarkably firm and commonly regular teeth. In the youthful these are in general white, but towards middle age they have lost their enamel and become black, or are worn down to the gums. The incisors of the lower jaw do not pass behind those of the upper, but meet edge to edge, so that by the time an individual arrives at maturity, the opposing surfaces of the eye and front teeth are perfectly flat, independently of the wear they are subjected to in every possible way to assist the hands. The expression of the countenance is one of habitual good-humour in the great majority of both sexes, but is a good deal marred in the men by wearing heavy lip ornaments.

The lower lip in early youth is perforated at each side opposite the eye-tooth; and a slender piece of ivory, smaller than a crow-quill, having one end broad and flat like the head of a nail or tack to rest against the gum, is inserted from within, to prevent the wound healing up. This is followed by others successively larger during a period of six months or longer, until the openings are sufficiently dilated to admit the lip ornaments or labrets. As the dilatation takes place in the direction of the fibres of the muscle surrounding the mouth, the incisions appears so very uniform as to lead one to suppose each tribe had a skilful operator for the purpose; this, however, is not the case, neither is there any ceremony attending the operation.

The labrets worn by the men are made of many different kinds of stone and even of coal, but the largest, most expensive, and most coveted, are each made of a flat circular piece of white stone, an inch and a half in diameter, the front surface of which is flat, and has cemented to it half of a large blue bead. The back surface is also flat, except at the centre, where a projection is left to fit the hole in the lip, with a broad expanded end to prevent it falling out, and so shaped as to lie in contact with the gum. It is surprising

how a man can face a breeze, however light, at 30° or 40° below
zero, with pieces of stone in contact with his face, yet it seems from
habit the unoccupied openings would be a greater inconvenience
than the labrets which fill them.

Their sight is remarkably acute, and seemed particularly so to us,
who often experienced a difficulty in estimating the true distance
and size of objects on the snow. Their hearing also is good, but we
doubt if it possesses the same degree of acuteness. Of the other
senses we have not been able to form an opinion.

While young the women are generally well-formed and good-look-
ing, having good eyes and teeth. To a few, who besides possessed
something of the Circassian cast of features, was attributed a certain
degree of brunette beauty. Their hands and feet are small, and
the former delicate in the young, but soon become rough and
coarse when the household cares devolve upon them. Their movements are awkward and ungainly, and though capable of making
long journeys on foot, it is almost painful to see many of them walk.
Unlike the men, they shuffle along commonly a little sideways,
with the toes turned inwards, stooping slightly forward as if carrying a burden; and their general appearance is not enhanced by the
coat being made large enough to accommodate a child on the back,
whilst the tight-fitting nether garment only serves to display the
deformity of their bow legs. Beyond the front view of the face,
they seem utterly regardless of cleanliness; and though careful in
arranging the beads in their hair, they seldom use a comb either for
comfort or tidiness. A sort of cleansing of the body generally is
occasionally practised, but it is far from deserving the name of
ablution. It is but fair to state that we believe they might be
easily taught habits of cleanliness, but these could be attended to
with the greatest difficulty, as they have no more water in the long
winter than is just sufficient for their drinking and cooking.
Around Michælowski, in Norton Sound, some of the women wear
cotton garments next the skin; and on bath days, after the people
of the Fort had done, they eagerly availed themselves of the
opportunity, when allowed, to wash both themselves and their
clothes.

The hair is worn parted in the middle from the back to the front,
and plaited on each side behind the ear into a roll, which hangs
down to the bosom and is wrapped round with small beads of
various colours. Length of hair generally accompanies softness of
its texture, and is considered a point of female beauty. The ears are,
with very few exceptions, pierced to support, with ivory or copper
hooks, four or five long strings of small beads suspended at a dis-

tance from the ends, which hang free, leaving the middle part to fall loosely across the breast. Not unfrequently the ends are long enough to be each fastened back in another loop to the hair behind the ears.

Fortunately for the appearance of the countenance it is not deformed by the perforations in the lip, but instead it is marked with three tattooed lines from the margin of the lower lip to the under surface of the chin. The middle one of these is rather more than half an inch broad, with a narrower one at a little distance on either side, diverging slightly downwards. The manner in which tattooing is performed is by pinching up the skin in the direction of the line required, and passing through it at short intervals a fine needle, in the eye of which is a small thread of sinew blackened with soot, as in ordinary sewing, except that the thread is pulled through at each stitch. The narrow line on each side is the result of one seam or series of stitches, but the middle one requires three or four such close together. It has been supposed that this operation is performed at a particular period when the girl verges into womanhood, and some of the natives profess that this is the case, but inquiry does not substantiate the supposition. A single line is frequently seen in mere children, and the three in very young girls, whilst a few are not marked until they seem almost full grown women, and have been called wives for a considerable time. The same irregularity exists with regard to the age at which the lip is perforated for labrets in boys, who as soon as they can take a seal or kill a wolf are entitled to have the operation performed. But, in truth, no rule obtains in either case: some, led by the force of example, submit to it early, and others delay it from shyness or timidity. A man is met with occasionally without holes for labrets, but a woman without the chin-marks we have never seen.

The men's dress is simple and convenient, consisting of a frock reaching nearly half-way to the knee, with a hood, and confined at the waist by a loose belt, having the tail of some animal attached to it behind, and breeches tying below the knee over long boots or moccassins, which also tie at the ankle. These garments are double, the inner being generally made of fawn-skin, and worn with the fur inwards, and the outer of the skin of the half or full-grown animal with the hair outwards. To make the hood set well to the face, a triangular slip of skin is necessary to be inserted on each side of the neck, with long points extending down the breast; and these pieces being usually white, form with the darker skin of the coat a contrast which readily catches the eye. Around the face is a fringe, frequently of wolf or wolverine-skin, on good coats, and the

skirt is hemmed with a narrow edging of a similar kind ; some have
also a border of white, with straps of the same colour on the arm
near the shoulder. There is commonly an ermine-skin, a feather,
or some such thing, which acts as a charm, attached to the back.
The skins of various other animals besides the deer, as the fox,
musk-rat, marten, dressed bird-skins, &c., are also used in making
coats. The breeches are also of deer-skin, or sometimes dog or seal-
skin, occasionally ornamented with a stripe of white down the outside or front of the thigh. The boots are most frequently of the
dark skin of the reindeer's legs, or this in alternate stripes with the
white skin of the belly, extending from below the knee to the ankle,
with soles of white dressed seal-skin, gathered in neatly around the
toes and heels, having within a cushion of whalebone scrapings or
dried grass, between them and the reindeer stockings, which are
next the feet. They are particular in the arrangement of the skins ;
thus the round spot of indurated skin on which the hair is stiffer
and whiter than that around it just below the hock of the animal is
always placed over the inside of the ankle-bone in men's mocassins
at Point Barrow, and over the outer in women's ; but they say the
reverse is the custom at Point Hope. Over these a pair of ankle-
boots of black seal-skin, dressed only so far as to remove the hair,
with soles of narwhal-skin, is worn on the ice. The hands are
protected by deer-skin mittens, with the hair inwards ; but for cold
weather and working on the ice, the thicker skin of the polar bear,
with the hair outwards, is preferred, as it is warmer and less liable
to injury from getting wet. The whole dress is roomy, particularly
the coat, which has the sleeves large enough to allow the hands to
be withdrawn, one of the greatest comforts that can be imagined in
cold weather. In winter a cloak of dark and white deer-skins is
worn over the shoulders, held on by a thong across the throat, and
gives the whole figure a very gay appearance. According as the
wind is in front or on one side, the cloak can be turned as a pro-
tection against it. The usual belt is made of the smaller wing-
feathers of ducks, after the plumes are torn off, partly sewed and
partly woven with small plaited cords of sinew, taking care to keep
the glossy back surface of the feathers outwards, and their ends,
which form the edges of the belt, are confined by a narrow binding
of skin. In some of these there is a checkered appearance, produced
by alternate rows of black and white feathers ; but the white tápsi,
or belt, is certainly the gayest. The pipe-bag on one side, and the
knife on the other, suspended to the girdle supporting the breeches,
may be considered part of the usual dress. For procuring fire, the
flint and steel is used in the North, and kept in a little bag hanging

round the neck; and in Kotzebue Sound the pipe-bag contains two pieces of dry wood, with a small bow for rotating the one rapidly while firmly pressed against the other until fire is produced. In the absence of these, two lumps of iron pyrites are used to strike fire upon tinder, made by rubbing the down taken from the seeds of plants with charcoal. The tobacco-bag, or " del-la-mai'-yu," is the constant companion of men, women, and even children, and is kept also at the inner belt.

In summer, as their occupations are more in boats, the dress is somewhat different. The feet and legs are incased in watertight scalskin boots, and an outside coat of the same material, or of whale-gut, covers the body ; or these are made all in one, with a drawing-string round the face. The least valuable skins are also used at this time, as they soon become soiled and filthy with blubber, becoming quite unfit for a second season.

It would be impossible to enumerate the varieties of dress we witnessed at the grand summer dance, when, among new skin coats, might be seen the clean white-cotton shirt and the greasy and tattered Guernsey frock, besides others made up of odds and ends, such as cotton or silk handkerchiefs procured at the ship, showing that they were bound by no rule as to dress on the occasion. On the head of every dancer, however, was a band supporting one, two, or three large eagle's feathers, which, together with a streak of black-lead, either in a diagonal line across or down one side of the face, gave them a more savage appearance than they usually exhibit. Many of these head-bands were made of the skin of the head and neck of some animal or bird, of which the nose or beak was retained to project from the middle of the forehead. The long beak of the great northern diver formed the most conspicuous of these ornaments. Another head-dress, which is looked upon with superstitious regard, and only worn when engaged in whaling, consists of a band of deer-skin ornamented with needlework, from which are suspended around the forehead and temples, in the form of a fringe, the front teeth of the im'-na, a sort of deer, which has been before mentioned as inhabiting the interior.

Snow-shoes are so seldom used in the North when the drifted snow presents a hard frozen surface to walk upon, that certainly not half a dozen pairs were in existence at Point Barrow at the time of our arrival, and those were of an inferior sort. Inland, and near Kotzebue Sound, where trees and underwood grow, the snow remains so soft it would be impossible to travel any distance in the winter without them. The most common one is two pieces of alder, about two feet and a half long, curved towards each other at the ends,

s

where they are bound together, and kept apart in the middle by two cross-pieces, each end of which is held in a mortice. Between the cross-pieces is stretched a stout thong, lengthwise and across, for the foot to rest upon, with another which first forms a loop to allow the toes to pass beneath ; this is carried round the back of the ankle to the opposite side of the foot, so as to sling the snow-shoe under the joint of the great toe. As the shoe is thus suspended at a point a little before its centre, the heel end trails lightly over the snow at each step, whilst the toe is raised over any slight unevenness in the way. Some are five feet long by fourteen inches wide, rounded and turned up at the toe, and pointed at the heel, neatly filled in before and behind the cross-bars with a network of sinew, or of a very small thong made from the skin of the small seal, nat'-sik.

The women's dress differs from the men's in the mocassins and breeches forming a single close-fitting garment tied round the waist, as well as in being more uniformly striped, and the coat in being longer, reaching to below the knees in a rounded flap before and behind. The back of the coat and the hood are also made large enough to contain a child, whose weight is chiefly sustained by the belt. For common use, and among the poorer people, the inner one is made of bird-skins, and among those who are better off, of deer-skin, and is plain. In winter, when out of doors, an outer coat of thick deer-skin is worn, and in summer a light one of the skins procured during the summer when the animal is changing its hair. For dress occasions, one is worn by those who can afford it which is made of patchwork, always according to one invariable plan as to the shape and principal seams ; but there is considerable variety allowed in the arrangement of the white and different shades of fawn-skins of which it is made, besides a countless multitude of strips and tufts of fir sewed to the back, shoulders, and front of the garment, producing always a pleasing effect, and indicating considerable industry on the part of the seamstress.

The woman's tápsi or belt is made from the skin of the wolverine's feet, with the claws directed downwards and placed at regular intervals. Near Kotzebue Sound a belt of a different kind is much in use, consisting of a piece of skin, of proper length, having the front teeth of the reindeer, adhering to the dried gum of the animal, stitched to it ; so that the second row of teeth overlies the sewing on the first, and so on, beginning at each end and joining at the middle. A belt of this description is about two and a half inches broad, and has from fifty to sixty rows of teeth. The other personal ornaments, besides the beads in the hair and ears, are rings of iron and copper for the wrists, and on dancing occasions their wealth is displayed in broad bands of small beads of different colours, arranged according to the taste of the wearer, attached by one end to the coat at the neck, and by the other to the middle of the front skirt. Large beads seem to be used only by the men, some of whom were vain enough to display them in strings round the head or hanging in front of the coat, and we remarked that no part of the materials procured from the ship was used as clothing by the women. Buttons were the only ornaments they seemed to adopt for the belt, and to fasten the beads in their hair.

Instead of a knife the women wear at the inner belt a needle-case, which is merely a narrow strip of skin in which the needles are stuck, with a tube of bone, ivory, or iron to slide down over them, and kept from slipping off the lower end by a knot or large bead. Their pipe is commonly smaller and lighter than the men's, and they do not carry it in a bag, but in the hand or inside the coat at the back; and the flint and steel is not so general with them, as their work is seldom out of doors except in company with the men. They have a singular habit of wearing only one mitten, protecting the other hand under the flap of the coat, or drawing it inside the sleeve, in preference to carrying a second.

The shape of the coat serves to distinguish the sex of children as soon as they are able to walk alone, but the woman's form of mocassins is used by boys until they are well grown.

The physical constitution of both sexes is strong, and they bear exposure during the coldest weather for many hours together without appearing inconvenienced, further than occasional frost-bites on the cheeks. They also show great endurance of fatigue during their journeys in the summer, particularly that part in which they require to drag the family boat, laden with their summer tent and all their moveables, on a sledge over the ice.

Extreme longevity is probably not unknown among them; but as they take no heed to number the years as they pass, they can form no guess of their own ages, invariably stating "they have many years." Judging altogether from appearance, a man whom we saw in the neighbourhood of Kotzebue Sound could not be less than eighty years of age. He had long been confined to his bed, and appeared quite in his dotage. There was another at Point Barrow, whose wrinkled face, silvery hair, toothless gums, and shrunk limbs indicated an age nothing short of seventy-five. This man died in the month of April 1853, and had paid a visit to the ship only a few days before, when his intellect seemed unimpaired, and his vision wonderfully acute for his time of life. There is another still alive, who is said to be a few years older.

Before offering any remarks on the character of these people, it should be premised that the subject is approached with great diffidence, lest we should give erroneous views respecting them ; for although we have resided two years within three miles of their largest settlement, we could never wholly divest ourselves of the feeling that we were looked upon by them as foreigners, if not intruders, who were more feared than trusted ; the more favourable points of their character were not therefore brought prominently before us ; whilst from being frequently annoyed by petty thefts, false reports, broken promises, and evasions, we perhaps too hastily concluded that thieving and lying were their natural characteristics, without attributing to them a single redeeming quality. Yet, as we became better acquainted, we found individuals of weight and influence among them, whose conduct seemed guided by a rude inward sense of honesty and truth, and whom it would be unfair to judge by a civilized standard, or to blame for yielding to temptations to them greater than we can conceive. A leaf of tobacco is a matter of small value, yet the end of it sticking from one's pocket amid a knot of natives at Nu-wuk, would be a greater temptation there, and would more surely be stolen than a handkerchief or a purse seen dangling from one's skirt in a London mob. And when the parental and filial duties are so carefully performed, it would be hard to deny the existence of even a spark of generosity.

In disposition they are good-humoured and cheerful, seemingly burdened by no care. Their feelings are lively but not lasting, and the temper frequently quick, but placable. Of their placable temper, an instance occurred in September 1852. An old man, of some consideration at Nu-wuk, had with his wife been alongside the ship, and in the crowd were refused admittance ; the woman also, by some accident, had received a blow on the head from an oar. By way of retaliation, a day or two afterwards he tried to send away our watering-party from a pond near the village ; and finding our men took little heed of him, he set about persuading his countrymen to expel the strangers "for stealing the water." Captain Maguire seeing the disturbed state of his feelings depicted in his countenance, advanced to meet him, and at once presented him with a needle. The man's embarrassment was extreme. Trifling as the present was, it flattered him out of more than half of his anger, and he dissipated the rest in a long talk, the people seating themselves in a ring, and requesting the captain and his companions to take a place in the centre, when the old man and his wife—his better half—

explained the bad treatment they had received at the ship. In the meantime the boat was laden, and the distribution of a little tobacco left a momentary impression that we were angels.

Their conjugal and parental affections are strong, the latter especially, whilst the children are still young; but beyond the sphere of their own family or but they appear to have no regard. The loss of a husband, a wife, or a child, makes no permanent deep impression, unless the bereavement leaves them destitute of the comforts they have been accustomed to; indeed, it is not rare to find a woman unable to give an accurate account of her children, including the dead; yet when their afflictions are brought to mind by inquiry, the cheerful smile leaves the face to be replaced by a look of sadness, and the tone of the voice becomes doleful. Under the real or pretended influence of grief, acts of violence are sometimes committed by the men, and thefts at the ship were occasionally said to be prompted by domestic sorrows. Though thankful at times for favours, they seldom offered any return, and gratitude beyond the hour is not to be looked for. Perhaps it is not too much to say that a free and disinterested gift is totally unknown among them. On making a present to a stranger, it was not uncommon to see him put on a look of incredulity, and repeatedly ask if it were really a gift.

They vied with each other for a long time in pilfering from the ship, whilst among themselves honesty seemed to prevail; but as we came to know them better, and were able to detect delinquents, our losses became fewer, and we learned that thefts from each other were not unfrequent, so that we arrived at the very unsatisfactory conclusion that it is the certainty of detection that prevents theft. Many articles, such as spears and other implements, are left exposed, and run no risk, as they would certainly be recognized by many others besides the owner; but when food, oil, tobacco, or such other things as would be difficult to identify, are concerned, the case is different. In the long passage leading to the winter hut, many articles are kept which could be easily taken unknown to the inmates; but during the day some neighbour would be sure to see the thief, or, if the deed were done in the night, his footmarks on the snow would tell the tale. It is in the stormy, dark nights the Nu-wuk burglar goes his rounds, trusting to the snow-drift to obliterate his footsteps. His visits are not unprovided against, for a trap is laid in most huts, not to catch the marauder, but to alarm and drive him away. This is affected by placing a board with a large wooden vessel on it in such a position, that both may fall on the slightest touch, thereby making sufficient noise to arouse the household, some of whom get up, re-adjust the trap, and retire again. We were also informed of instances as they occurred of stealing from each other seals left on the ice, and in one case a net was taken up and carried off to Cape Smyth.

It is almost natural to expect that falsehood should follow to conceal theft, and we found it here accordingly. To invent stories disparaging to others was a practice some addicted themselves to without any conceivable motive, and the women backbite each other and talk scandal very freely. Their confidence in our honesty soon became unbounded, and goods brought to the ship and not disposed of were frequently left behind; yet though they knew our engagements would be fulfilled, when a bargain was made they appeared uneasy until the payment was effected. Selfish gratification at the present moment is all they seem to live for, and no promise of a reward, however great, would induce them to deviate from their usual life for any continued period.

If they do not possess courage of a daring character, they have given us no reason to look upon them as cowards. When the crew of Mr. Shedden's vessel, the *Nancy Dawson,* landed on the ice to shoot birds, the handful of men whose tents were in the neighbourhood advanced, bow in hand, to meet them and drive them back. Some of these men have since explained, that fearing the guns, they thought it better to oppose the landing of the strangers than trust them on shore, before knowing them to be friends; adding, that " Mr. Martin was a good man, who said they were friends, and made the ship's people put away their guns." After committing a robbery at our storehouse, they attempted to direct attention to the Cape Smyth people as the thieves, although the track left by dragging some sails had been followed to near Nu-wuk. When this was pointed out, and a threat made to send an armed force to recover the stolen property, they turned out to the number of eighty men, with bows and spears, and advanced within musket shot of the ship, rather than stand a siege in their own dwellings. We have learned enough from them to believe they at first looked upon us as a contemptible few whom they could easily overcome, and certainly would have attempted it but for fear of the firearms; but since then, they have gone to the opposite extreme, and invested us with greater powers than we really possess. On trifling occasions some of them have shown a degree of obstinacy which renders it probable, that if once engaged in a fight they would not readily give in, at least if there was anything like equality of weapons; and, under any circumstances, they might be expected to defend their homes to the last extremity.

Being in the habit of making frequent journeys of four or five days without taking more than two days' provisions, they appear to rely on the kindness of others as they pass; and as this is perhaps never denied, hospitality to strangers may be esteemed a duty. We are of opinion, however, this has its limits. A man of good name would have no difficulty in procuring food and shelter while travelling through any part of his country, as, where he ceased to be known by his own reputation, he would be accepted as a guest on mentioning the name of his last entertainer; and we have never entered a strange hut without inquiry being made as to what sort of food we used, and generally some of their best was set before us, or an apology made that they had nothing to offer which we would relish. But an Eskimo never undertakes a distant journey unless he well knows the people he is going among, or he goes in company with others on whom he can depend for a welcome. In a society so large as that at Point Barrow, it is impossible that different families should be at all times totally independent of each other, and the successful hunter of to-day lends to his neighbour, who, when the luck turns, repays the favour; but dealings of this kind are practised no more than necessity requires. A man returned during the hunting-time to the village, and his own hut being closed, he lived with a relative for four or five days; in return for which, when the season was over, that relative and some of his family spent a whole day in the other's hut, where they were entertained with reindeer-flesh, which was then very scarce.

For the tender solicitude with which their own infancy and childhood have been tended, in the treatment of their aged and infirm parents they make a return which redounds to their credit, for they not only give them food and clothing, sharing with them every comfort they possess, but on their longest and most fatiguing journeys make provision for their easy conveyance. In this way we witnessed among the people of fourteen summer tents and as many boats, one crippled old man, a blind and helpless old woman, two grown-up women with sprained ankles, and one other old invalid, besides children of various ages, carried by their respective families, who had done the same for the two first during many successive summers. Here, again, the tie of kindred dictates the duty, and we fear it would go hard with the childless. When a man dies, his next of kin supports his widow; or if unprovided already, he may make her his wife, unless he allows her to be taken by a stranger. Orphan children are provided for in the same way, and adoption is so frequent among them that it becomes almost impossible to trace relationship; this is, however, of no importance,

as the adopted takes the place of a real child, and performs his
duties towards his benefactors as if for his own parents. Grief is
sometimes made the excuse for violence, but it is also assuaged in a
nobler manner by adopting the children of the deceased; or a
stranger's orphan, to whom the name of the lost one is given. In
this manner O-mig-a-loon, the principal man at Point Barrow, the
same who followed and annoyed Captain Pullen at Point Berens,
adopted an Indian infant which fell into his hands by accident
while grieving for his father, then recently dead, whose name the
youth now bears. We have never heard of the sick or aged being
left to perish, though at Icy Cape we saw a woman lying dead,
in a hut, who had been subject to bad treatment, as evidenced by
the bruises on her face. Within her reach were placed food and
water, which we are willing to look upon as proofs that it was not
intended she should die of starvation. One instance of infanticide
came within our knowledge during the last winter ; but a child,
they say, is only destroyed when afflicted with disease of a fatal
tendency, or, in scarce seasons, when one or both parents die. In
the case alluded to both these conditions were present. They state
that children are rarely put to death at Nu-wuk, though frequently
in the inland regions ; as if by pointing out its greater frequency
there they palliated the crime among themselves.

Having but little food of a nature adapted to supply the place of
milk, it is no unusual thing to see a boy of four or five years old
take the breast ; and the indulgence with which children are
treated is attributable in some degree to the difficulty in rearing
them. We have seen a child of four years old demand a chew of
tobacco from his father, and, not receiving it immediately, strike
him a severe blow on the face with a piece of wood, without giving
offence. It is not improbable that such indulgence should have a
permanent effect on the temper and character of the people. The
children fight with and bully each other in their play, but among
the grown-up men or women we have never seen anything approaching a quarrel ; and, as a general rule, they are particularly careful
not to say anything displeasing in each other's presence. If a man
gets angry or out of temper, the others, even his nearest friends,
keep out of his way, trusting to his recovery in a short time.
Whenever we have met them at a distance from the ship in small
parties, they have proved tractable and willing to assist when required ; but when the numbers were large they were mischievous
bullies, threatened to use their knives on the slightest provocation,
and, instead of giving assistance, would rather throw impediments
in our way. We hardly think them likely to commit wanton cruelty,

or to shed human blood without a strong motive, yet we would be unwilling to trust to the humanity of a people whose cupidity is easily excited, and who are accustomed to no restraint save their own free will. When murder is committed, as it sometimes is, it is in retaliation for injury, real or fancied; and then the victim is stolen upon while asleep and overpowered by numbers, or he receives his death-wound unawares from some one behind him.

In point of intelligence, some exhibit considerable capacity, and in general they are observant and shrewd. As a people, they are very communicative, those of most consideration being generally most silent; and wisdom is commonly imputed to those who talk least. They possess great curiosity, and are chiefly attracted by whatever might be useful to themselves. In this way a gun would be a study they seemed never to tire of, particularly the lock; and the blacksmith when working at the forge was, perhaps, as great an attraction as there was on board the ship. They soon began to appreciate prints and drawings, and latterly often borrowed books of plates to amuse them at home, always taking great care of them and returning them in good order. When shown the construction of a pair of bellows, a few appeared to perceive and admire the mechanism at once, whilst to many it remained quite a mystery to the end. They were totally unable to comprehend how the sounds were produced from a flute, and it was highly amusing to see one of the most intelligent amongst them, who fancied there was some trick practised, examine the fingers and lips of the musician to find out the deceit. Every article that fell under their notice became the subject of inquiry as to what were its uses, the material it was made from, how it was manufactured, and if it pleased them much, the name of the maker. At first they exhibited some caution in receiving information, and went slyly from one to another asking the same questions; but latterly they ceased to do so. A perfect stranger, especially if young, and allowed to roam at large about the ship, would in a short time be able to name almost every one on board, but in a way hardly recognisable. One boy at the end of six months could count on his fingers as far as ten, mastering the letter *f* in four and five tolerably, but still with great effort; and learned a few other words. A number of others tried at first to follow his example, without success; and it was remarked that "pease-soup" was the only English word generally known and distinctly pronounced. The majority have a strong sense of the ludicrous, and readily observe personal peculiarities, which they will afterwards describe with great zest. Some of them are tolerable mimics, and their efforts are sure to meet with applause, especially

when the subject is a stranger; but among themselves they are
very discreet in the exercise of this faculty. A few of the men
showed some quickness in interpreting the drift of our inquiries
respecting their superstitions and usages; but for the insight we
gained of these we were usually indebted to the women—especially
the younger ones, who, besides being more communicative, displayed
more readiness in this respect—for the first information, which,
being afterwards confirmed by the older men, served as a clue to
guide farther inquiry.

A man seems to have unlimited authority in his own hut, but as,
with few exceptions, his rule is mild, the domestic and social position of the women is one of comfort and enjoyment. As there is no
affected dignity or importance in the men, they do not make mere
slaves and drudges of the women; on the contrary, they endure
their full share of fatigue and hardship in the coldest season of the
year, only calling in the assistance of the women if too wearied
themselves to bring in the fruits of their own industry and patience;
and at other seasons the women appear to think it a privation not
to share the labours of the men. A woman's ordinary occupations
are sewing, the preparation of skins for making and mending,
cooking, and the general care of the supplies of provisions. Occasionally in the winter she is sent out on the ice for a seal which her
husband has taken, to which she is guided by his foot-marks; and
in spring and summer she takes her place in the boat, if required.
Seniority gives precedence when there are several women in one
hut, and the sway of the elder in the direction of everything connected with her duties seems never disputed. In the superinten-
dence of household affairs the active mother of the master of a hut
or of his wife must be a great acquisition to his family, from her experience and from the care and interest she displays in their management; and, as her natural desire is to see her children happy around
her, she exerts herself to promote their well-being and harmony.

It is said by themselves that the women are very continent before
marriage, as well as faithful afterwards to their husbands; and this
seems to a certain extent true. In their conduct towards strangers,
the elderly women frequently exhibit a shameless want of modesty,
and the men an equally shameless indifference, except for the reward of their partner's frailty. In the neighbourhood of Port
Clarence this is less the case than farther north, whilst on the Island
of St. Lawrence it is, perhaps, more so than on any part of the coast.

The state of wedlock is entered at a variable time, but seldom in
extreme youth, unless as a convenience to the elders, who desire
an addition to the household. The usual case is, that as soon as the

young man desires a partner, and is able to support one, his mother
selects a girl according to her judgment or fancy, and invites her
to the hut, where she first takes the part of a " kir-gak " or servant,
having all the cooking and other kitchen duties to perform during
the day, and returns to her home at night. If her conduct prove
satisfactory, she is further invited to become a member of the family,
and this being agreed to, the old people present her with a new
suit of clothes. The intimacy between the young couple appears to
spring up very gradually, and a great many changes take place before
a permanent choice is made. Obedience seems to be the great
virtue required, and is enforced by blows when necessary, until the
man's authority is established. In the ordinary course of events
life runs smoothly enough, and is only checked by a few lover's
quarrels or fits of sulkiness; but it occasionally happens that the
husband finds his regard unrequited, and he either trusts to time to
overcome her indifference, keeping a strict watch over her conduct,
or he treats her with severity. The consequence of this is her return to her friends, whither he may follow and drag her back to his
hut. Repeated occurrences of this kind may take place and end in
permanent harmony ; but if his treatment has been cruel, which it
seldom is to their view, and her relatives not interested in enforcing
the union, she is taken back and protected from his farther violence.
We have been assured it sometimes happens that several men
entertain a passion for the same woman, the result of which is a
fight with bows and arrows, ending in the death of some of the
aspirants, and she falls to the lot of the victor. A man of mature
years chooses a wife for himself, and fetches her home, frequently, to
all appearance, much against her will ; but she manages in a wonderfully short time to get reconciled to her lot. A union once apparently settled between parties grown-up is rarely dissolved ; though
we have seen a woman and her child residing with her relatives,
having been deserted by her husband, for what reason could not be
ascertained. The woman's property, consisting of her beads and
other ornaments, her needle-case, knife, &c., are considered her own ;
and if a separation takes place, the clothes and presents are returned,
and she merely takes away with her whatever she has brought.
Unless she has proved an untameable shrew she need not be appre-
hensive of remaining long single, as the proportion of males to
females in the population is more than eight to seven, besides
which several of the leading men have each two wives.

Bigamy is evidently looked upon as a sign of wealth, and is in
many instances analogous to the adoption of children. Thus, if a
man is a trader and well off, he may require the assistance of

another woman to work up his peltry into coats for the next market ; or his wife may be nursing, and cannot well perform all the duties that usually devolve upon the mistress of a large establishment. Under such circumstances he may take home as an additional help-mate some elderly widow, and both parties will be benefited by the arrangement. This is, however, not always the motive, and no little jealousy is sometimes excited by the introduction of a younger and better-looking woman to the establishment. The practice is, after all, not very common, as only four men out of a population of near 290 at Point Barrow had each two wives. There were four also at Cape Smyth, where the population is smaller, and several at Point Hope. At the latter place one was particularly mentioned as having no less than five wives ; and although it is the only instance of polygamy we heard of, it serves to show that custom has put no limit to the number of wives a native of this country may have.

The age at which the women are married is probably in general fifteen to sixteen. They do not commonly bear children before twenty ; and there is usually an interval of four years or more between the births. They relate, apparently with little hope of being believed, that some years ago a woman at Cape Smyth had two children at one birth. For one woman to have borne seven children is a rare case, and for five to live to maturity still more rare. If any one in the ship were stated to be the ninth or tenth child of one family it excited their astonishment, and if to this it were added that seven or eight of them were still alive, they became incredulous. A couple is seldom met with more than three of a family, though inquiry may elicit the information that one or several "sleep on the earth." From this, and the great care and in-dulgence with which those of tender years are treated, it may be inferred that the greatest mortality takes place under the fifth year, but it does not appear that there is any particular form of disease to which they are, before this age, peculiarly liable ; the condition of the mother, however, according as the season is one of abundance or scarcity, has by their own account a material influence on the health of the offspring. During first pregnancy great solicitude has been observed on the part of the husband for his wife, although there is no reason to believe childbirth anything but easy. In the particular instance alluded to, from the delicate appearance of the woman it was fancied that every precaution was taken to guard against premature labour, three cases of which came under notice in the last winter.

Previous to proceeding farther with the usages and occupations of these people, it will be well to give some idea of their habitations.

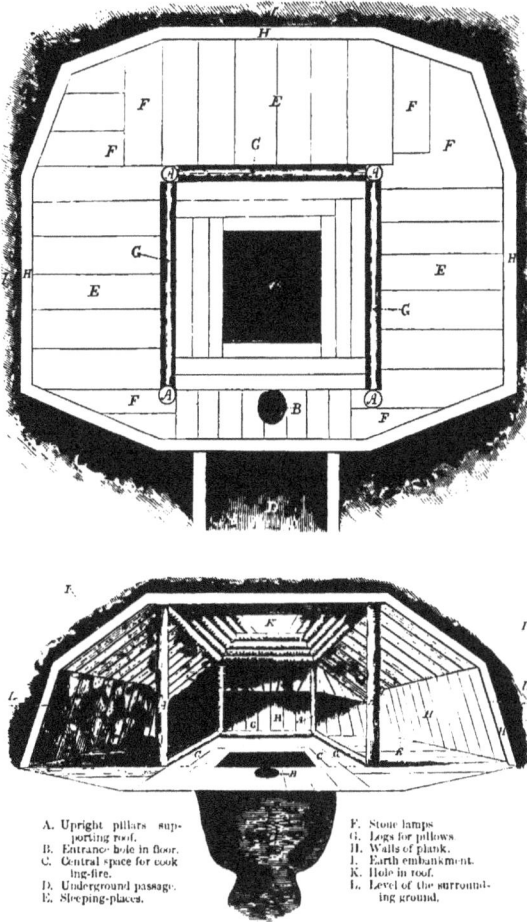

A. Upright pillars sup-
   porting roof.
B. Entrance hole in floor.
C. Central space for cook
   ing-fire.
D. Underground passage.
E. Sleeping-places.

F. Stone lamps
G. Logs for pillows
H. Walls of plank.
I. Earth embankment.
K. Hole in roof.
L. Level of the surround-
   ing ground.

GROUND-PLAN AND INTERIOR OF ESKIMO WINTER-HUT AT HOTHAM INLET.

The winter huts at Point Barrow are not placed with any regard to order or regularity, but form a scattered and confused group of grassy mounds, each of which generally covers two separate dwellings, with separate entrances; some, however, are single, and a few are threefold. Behind each are placed a number of tall posts of driftwood, with others fastened across them, to form a stage on which are kept small boats or kaiaks, skins, food, &c., above the height to which the snow may be expected to bank up in the winter, and beyond the reach of dogs. These posts show out very plainly against the horizon in the winter, when everything beneath is covered with snow, and in all seasons may be seen at a considerable distance, long before the huts themselves become visible. The entrance to each hut is from the south by a square opening at one end of the roof of a passage twenty-five feet long, and has a slab of ice or other substance of convenient shape to close it at pleasure. The passage, which is at first six feet high, descends gradually until about five feet below the surface of the ground, becoming low and narrow before it terminates beneath the floor of the hut. Near its middle on one side branches off a recess, ten to twelve feet long, with a conical roof open at the top, forming an apartment which serves as a cook-house, and on the other is commonly enough a similar place, used as a store or clothes' room. The "iglu" or dwelling-place is entered by a round aperture in the floor on the side next the passage, and is a single chamber of a square form, varying in size from twelve to fourteen feet from north to south, by eight to ten from east to west. The roof has a double slope of unequal extent, that on the south side being the larger, with a square opening or window, covered with a transparent membrane stretched into a dome-shape by two pieces of whalebone arched from corner to corner, and is generally a little more than five feet high under the ridge. The smaller part of the roof has between it and the floor a bench, on which a part of the family sleep at night, and sit or lounge during the day. The walls are of stout planks, placed perpendicularly, close at the seams and carefully smoothed on the inside; the floor and sleeping-bench are the same, whilst overhead are small rounded beams, also smoothed and scraped, sustaining the weight of the earth heaped on top. As the bench and the sleeping-place beneath do not in many instances exceed four feet from the wall to the cross-beam at the edge, which serves as a pillow, the occupants cannot be supposed to lie at full length, but this limited extent of the bed-place gives greater space in the other part of the hut, which is thus left nearly square, and is generally occupied by the women sewing or performing other household duties. The entrance and bed-place are at opposite ends; and on either hand is an oil-burner or fireplace, having a slender rack of wood suspended over it, on which articles of clothing are placed to dry, also a block of snow to melt and drip into a large wooden vessel. Beneath the last again are other vessels for different purposes, some of them frequently containing skins to undergo preparation for being dressed. These vessels are each made of a thin board of the breadth required, bent into the form of a hoop, and the ends sewed together neatly with strips of whalebone, the bottom being retained in its place by a score like the end of an ordinary cask. The oil-burner is the most curious, if not the most important piece of furniture in the establishment. It is purchased ready made from the eastern Eskimo, who procure it from a more distant people. It is a flat stone of peculiar shape, three to four and a half feet long, and four inches thick, pointed at the ends by the union of the two unequally convex sides somewhat like the gibbous moon. The upper surface is hollowed to the depth of three-quarters of an inch to contain the oil, leaving merely a thin lip all round, and several narrow ridges dividing the hollow part both lengthwise and transversely. It is placed on two horizontal pieces of wood fixed in the side of the hut, about a foot from the floor, with the most convex side towards the wall, the other being that where a broad flame of any extent required is sustained from whale or seal-oil by means of dry moss for wicks. When the length of one side of a lamp of this description is considered, it will readily be conceived that not only a good light but also a great deal of heat may be produced, so that the temperature of a hut is seldom below 70° of Fahr., though we have hardly ever seen a flame of more than a foot in extent; and as great care is taken to keep it trimmed, no offensive degree of smoke arises, though the olfactories are saluted on first entering by a combination of scents anything but agreeable. Ventilation is not altogether neglected, as there is near the middle of the roof a hole in which a funnel of stiff hide is inserted to carry off the vitiated air from the interior of the hut. When the place is much crowded or the temperature too high, a corner of the membrane can be raised; but we have seen it more speedily effected by the master of a house at Nu-wuk, in his impatience to contribute to our comfort, by making an incision with his knife through the middle of it—a proceeding which did not seem to be entirely approved of by his wife, to whose lot it would doubtless fall to repair it.

Such are the usual habitations on the coast of the Arctic Sea; but there are also others of a greater extent and different form, one of

which near the entrance of Hotham Inlet, Kotzebue Sound, is worth
mentioning, more particularly as it bears some resemblance to one
described by Sir John Richardson, on the east side of the Mackenzie
River. The outside did not differ.in appearance from the others,
except in size, as indeed they were all pretty well covered with
snow, but the interior was in shape something like three sides of a
cross, twenty feet by sixteen, with a roof sloping down on all sides,
like that of a verandah, from a square framework in the centre,
supported by four straight pillars, one at each corner, seven feet
high and eight feet apart. The quadrangular space in the centre was
covered with loose boards, which were removed when the fire was
required for cooking. It was bounded by logs stretching between
the bases of the pillars, and rounded on the upper surface to rest
the head upon during sleep, and had above it the usual square
aperture answering alternately the purpose of a chimney and a
window. Three sides of the house formed as many recesses, five
and a half feet from the log stretching between the pillars to the
walls, and were occupied at the time of our visit by six families,
each family having their own lamp in the intervals between the
recesses. The fourth side was only two feet deep, and left
space for little more than the entrance-hole in the floor and a
few household utensils. The walls were only three feet high,
and inclined slightly inwards the better to support the sloping roof,
which, like them and the flooring of the recesses, was made of
boards nearly two feet broad, quite smooth and neatly joined.
The whole building was remarkable for the regularity of the form
of the interior, and for the mechanical skill displayed in the work-
manship. Huts of this description may be looked upon as a combination of several, each recess representing a separate establishment, united in this form for mutual convenience, and are used
where driftwood is abundant, the large cooking-fire in the middle
of the building imparting its warmth to all around. But the
rushing down of cold air, and the smoke not always ascending,
proved sources of greater discomfort to us whenever we visited
them than the close atmosphere of those in which oil only is
burned.

A modification of the last form, built of undressed timber, and
sometimes of very small dimensions, with two recesses opposite
each other, and raised about a foot above the middle space, is very
common on the shores of Kotzebue Sound; but on the rivers, where
trees grow, structures of a less permanent kind are erected. Then
the smaller trees are felled, cut to the length required, and split;
then laid inclining inwards in a pyramidal form, towards a rude

square frame in the centre, supported by two or more upright posts. Upon these the smaller branches of the felled trees are placed, and the whole, except the aperture at the top and a small opening on one side, is covered with earth or only snow. The entrance is formed of a low porch, having a black bear-skin hanging in front, leading to a hole close to the ground, through which an unpractised person can hardly creep, farther protected from the breeze by a flap of deer-skin on the inside. In the hilly districts, near the source of the Spafaroif River, this sort of snow-covered hut was in use, and the inland tribes on the Nu-na-tak, are described as living in dwellings of a similar kind, constructed of small wood, probably built afresh every year, and not always in the same locality. A stranger approaching a village of this description, if the numerous footmarks happened to be obliterated by a recent drift or fall of snow, might readily pass by unconscious of its existence, unless he happened to catch a glimpse of the black bear-skin doors, which are all turned in the one direction.

Snow or ice huts are seldom used except for short intervals, and they are then made very small, consisting of two chambers, the outer one of which serves as a cook-house, and is entered from above by an opening closed at pleasure by a slab of snow. The communication between this and the inner one is by a passage close to the floor, no larger than necessary for one person to creep through. The roof of the inner apartment is about five feet high, with a window facing the south, having beneath it a small lamp and rack for drying clothes; and on one side the snow is raised two feet from the ground, and covered with boards, on which the skins are laid to form the bed.

In fixed settlements, like those of Point Barrow or Cape Smyth, there are other buildings which seem public, though nominally the property of some of the more wealthy men. In the former of these places there are two still in existence, and in the latter three. The largest is at Nu-wuk, and is eighteen feet by fourteen, built of planks stuck upright in the ground, and the crevices filled up with moss. The roof is similar to that of the other huts, only higher, and there is no sleeping bench within, but a low seat all round the four walls. It has the usual subterranean passage for entrance, but the window in the roof is often used as a door. Unlike the other huts, they are placed on the highest ground, and are readily distinguished by not being built around, or covered with earth. They are altogether constructed with little care, and evidently for only occasional use. A house of this description is called a Kar-ri-gi, and used by the men to assemble in for the purpose of

T

dancing, in which the women join, for working, conversing and idling, whilst the boys are unconsciously learning the customs and imbibing the sentiments of their elders.

In summer they live in conical shaped tents of deer or seal-skins, according as they are inland or coast people. Four or five poles, from twelve to thirteen feet long, slung together by a stout thong passing through holes in their tops, are spread out to the proper size, and within them, at a mark on each, about six feet from the ground, a large hoop is fastened. Smaller poles are then placed between the others in a circle on the ground, and leaning against the hoop to complete the frame of the tent. The skins are in two parts, each having a long corner sewed into a sort of pocket to fit the top of the long poles, over which one is placed above the other from opposite sides, so as to surround the whole framework, and allow the edges of one set of skins to overlap those of the other, and be secured by a few thongs. A large flap is sometimes cut in one side to form a window, fitted with a transparent membrane, over which the flap of skin may be replaced as a blind during sleeping-time. A tent of this kind is called a " tu'-pak," and makes a very comfortable summer abode, one side of which can be kept open to any extent, according to the weather: it is easily transported, and may be set up or taken down in an incredibly short time.

Commencing with the first new moon after the freezing-over of Elson Bay, which took place on the 24th of September, 1852, and on the 16th of September, 1853, the Point Barrow people divide the year into four seasons, which they call O'-ki-ak, including October, November, and December; O'-ki-ok, January, February, and March; O-pen-rak'-sak, April, May, and part of June; and O-pen-rak', the remaining part of June, together with July, August, and September. The successive moons, to the number of twelve, are also named by them, evidently in reference to their own occupations, to the phenomena observable in the season itself, or in animals, such as their migrations, &c., though we have been able to make out the precise meaning of only a few of them. These vary a little in different localities; but the setting-in of the winter being taken as the beginning of the year in all parts of the country, and the summer moons being but little noticed, no confusion seems to result. Taking them as they occurred in the last season, 1853–4 each tad'-kak or moon was given us as follows.

    I. 1853, Oct.   2, Shud'-le-wing, sewing.

    II.   „   Nov.   1, Shud'-le-wing ai-pa, sewing.

    III.   „   Nov.   30, Kai-wig'-win, rejoicing.

IV. „ Dec. 30, Au-lak'-to-win, departing (to hunt the
             reindeer).
V. 1854, Jan. 28. Ir'-ra shn'-ga-run sha-ke-nat'-si-a, great
             cold (and) new sun.
VI. „ Feb. 27, E-sek-si-la' wing.
VII. „ Mar. 28, Kat-tet-a'-wak, returning for whale (from
             hunting ground).
VIII. „ April 27, Ka-wait-piv'-i-en, birds arrive.
IX. „ May 26, Ka-wai-a-niv'-i-en, birds hatched.
X. „ June 25, Ka-wai'-lan pa-yan-ra'-wi-en, (young) birds
             fledged.
XI. „ July 25, A-mi-rak'-si-win.
XII. „ Aug. 23, It-ko-wak'-to-win.

As the new moon of September falls on the 21st of the month, it
will require an early setting-in of the winter to make that the first
moon of the next year.

For denoting time they also have expressions equivalent to yesterday, to-day, to-morrow, morning, afternoon, evening, &c., but
these are not by any means precise ; and in speaking of events a
year or more past, they use two terms, ai-pa'-ne, which seems
properly to mean two years ago (ai'-pa, two), but may be as readily
applied to twenty ; and al-ra'-ne, in the olden time, which is exceedingly indefinite. They have frequently declared that they keep
no account of the years as they roll, and " never number them, as
they do not write like us," so that it is next to impossible to get anything like exact dates from them. In describing the direction of
any distant place they are equally vague, using the term a-wa'-ne,
westward, or along the coast towards Icy Cape or Point Hope ;
ka-wa'-ne, eastward, or towards the Colville or Mackenzie rivers ;
pa-ne, south, or landward ; and u-na'-ne, north or seaward.

The seasons, as mentioned above, seem to guide them almost ᴛ
instinctively in their different occupations ; and it will not perhaps
be amiss to enumerate the principal ones which employ their time
throughout the year.

In the month of September they have almost all assembled at
the winter huts, amongst which they pitch their seal-skin tents,
living in them in preference to the yet damp underground ig-lu's,
and are constantly on the look-out for whales, killing also a few
walrus, bears, and seals, until the winter has fairly set in, and the
sea become shut up with ice, which generally takes place about
the middle of October. During this time most of the women
remain in comparative idleness at home, "as it is not good for them
to sew while the men are out in the boats;" but so soon are these are

laid-up for the winter, the sewing, together with cleaning the skins, commences, and is most industriously carried on for two months following. The men are now also engaged in setting nets under the ice for seals, in catching small fish with hook and line through holes in the ice, or in preparing implements used at other seasons. As midwinter approaches, the new dresses are completed, and about ten days at this season are spent in enjoyments, chiefly dancing in the kar-ri-gi, every one appearing in his or her best attire. This time of the year being one in which hunting or fishing cannot well be attended to, and no indoor work remaining to be performed, is perhaps sufficient reason why it should be chosen for festivities in the high latitude of Point Barrow, when the sun is not visible for about seventy days ; but it may not equally explain the prevalence of the same custom about the same period in Kotzebue Sound, lat. 66°, when the reindeer might be successfully pursued throughout the winter, the people then collecting from many miles around, to hold a festival in the neighbourhood of Cape Krusenstern. The amusements being concluded, a few set out early in January ; but it is later when the larger parties take their departure for the land in search of deer, scattering themselves over the flat ground at a variable distance of three to eight or ten days' journey from the village, and hollowing out dwellings in the deep snow-drift under the banks of the rivers, through the ice of which they make holes for catching fish by nets and for obtaining a supply of water. This occupies the majority of the people until April, the few who remain at home receiving supplies from time to time, besides spearing a few seals by watching for them as they come to breathe through the cracks in the ice ; or, if it is not in a favourable state for this near the shore, they make snow-houses to live in among the grounded masses in the offing. Having brought home the spoils of the chase, in the end of April they commence preparing their boats for launching and the implements used in capturing the whale, which gives employment for the men. The women are now also busily engaged in making watertight seal-skin boots and other articles of dress appropriate for summer wear. Towards the end of May, birds, chiefly eider and king-ducks, engage much attention from the whole population as they pass over the village northward, in rapidly succeeding flights of one to two hundred birds, alternately male and female. The whales having disappeared and the birds passed, a short interval is allowed to prepare dresses for another festival, which takes place in the end of June, and occupies six or eight days, when the dancing is performed in the open air. Early in July more than one-third of the community take their

departure in a body to the eastward, to make the long journey to
Colville River, and to Barter Point, many of the others following in
small parties to scatter themselves over the land in search of deer,
and over the lakes and rivers for birds and fish. About one-fourth
of the population remains at the village, catching abundance of
small seals, but chiefly looking out for those of a larger size, and
walrus, until the whales re-appear in the end of August, soon after
which, most of the travellers return from their wanderings to commence another year. At midsummer, when the sun has been some
time above the horizon, the snow becomes soft and the rivers begin
to flow, so that travelling or the pursuit of game is too fatiguing to
be successfully carried on; this season, therefore, like midwinter,
becomes necessarily one of comparative idleness, or is only spent in
amusement.

Such is a brief sketch of the ordinary annual routine of the occupations of the Eskimo of Point Barrow; but it is to be remarked
that unusual success or the reverse in hunting or fishing, more
especially as regards the whale, must always modify it in a great
degree. Thus, in 1852, no less than seventeen whales were said to
have been taken, sufficient to afford the poorest and most im-
provident abundance of food and fuel for the winter; and in the
succeeding spring, out of their superabundance of deer, a very considerable number was brought to the ship for barter; whilst, in 1853,
only seven whales, and those mostly small ones, were killed, giving
rise to such want of the necessaries of life in the last winter that
many families were obliged to use the decayed flesh and blubber of
a dead whale which had been stranded on Cooper's Island, about
twenty-five miles distant, more than two years before, and had
remained up to this time neglected. But even this resource failed
them, and many, as has been before mentioned, perished of famine.
In the former year, at midwinter, feasting and dancing were constant
for nearly a fortnight, and during October, November and December,
the number of seals offered for sale at the Plover was very great;
but in the latter they had none of these amusements, at least in
public, as they had not oil enough to spare for warming and lighting
up the dance-huts, and up to July only a few scraps of seal were
brought to the ship. The want of oil also prevented some of the
most wealthy men from going to hunt the deer in the winter; and
consequently none but a few pounds of venison were brought to
the ship for barter, the supply being hardly adequate to their own
wants.

From some of the more intelligent men, it appears that they
consider the last season one of uncommon privation, and that of

1852-3 was one of unusual abundance. Tracing back the years on the fingers, with some patience, it could be made out that in 1851-2 whales abounded, in 1850-1 the narwhal supplied the place of whale, giving them plenty of food and skins for covering their boats. 1848-9 was one of scarcity, as was also 1843-4. This, so far as it may be depended on, makes three successive fifth years to be seasons of unusual hardship. In 1837, Mr. T. Simpson remarked the number of fresh graves on Point Barrow, but no satisfactory account of the season preceding that could be obtained, and it was too remote to be recalled with anything approaching certainty by even those who remembered that gentleman's visit.

Having cleared out most of the furniture from the ig'-lu, and filled up the window with pieces of timber and other lumber placed on their ends, so as also to obstruct the entrance-hole in the floor, the um'-i-ak or large boat is put upon a sledge, u'-ni-ok, when it is secured by a few cords or thongs, and in it are stowed the summer tent with all its furniture, the baggage of the whole family, the children and old people, together with the kayaks or canoes, and all their fittings belonging to the men and boys of the party, making a very considerable weight to drag. On a low sledge, ka-mó-tik, of a stouter structure, are generally carried their seal-skins, filled with oil for barter. The party consists on the average of six persons, four of whom are generally all who can drag, and are distributed, three to the large sledge, and one to the ka-mó-tik. If they possess dogs, these are distributed also to assist where most required, and there appears to be as much care taken as possible to adapt the load to the strength of each individual. The ice at this season is much decayed and uneven from the formation of pools on its surface, and the labour of dragging a heavy load on a sledge is very great; but, fortunately for them, it seldom lasts more than four or five days, during which they appear to travel at the rate of ten miles a day. Fourteen parties, with as many boats (the aggregate number of souls being seventy-four), passed the ship in this way on the 3d of July last, which is four days earlier than in the preceding summer. On the fourth day they arrive at Dease Inlet, which, from the rivers flowing into it, is then a sheet of water, and the mode of transport is reversed, the sledge being now carried in the ú-mi-ak, and the small boats towed. In favourable seasons the journey may be continued by paddling or tracking the boat along the shore, between which and the ice there is generally a narrow lane of water, until they arrive at Smith's Bay. Here the laborious part of their journey is sure to end ; the sledges are left behind, and to make room in the large boat for the oil-skins, the men get into their kayaks. They enter a river which conducts them to a lake, or rather series of lakes, and descend another stream which joins the sea in Harrison Bay, within a day's journey and a half of the Colville. Whilst passing these streams and lakes they are enabled to supply themselves abundantly with fish of large size by nets ; a few birds are also taken, and occasionally a deer. About the eleventh day they encamp on a small island, within half a day's journey of the bartering place, and the different parties probably wait for each other there to enter the river in company.

The Colville River is described as having four mouths, the western of which is very shallow, but the second is a good deep channel, and is therefore followed until they get into the undivided stream, on the left or west bank of which they see the tents of their friends, the Nu-na-tang'-meun. Six, eight, or ten days, for precise numbers could not be obtained, are spent in bartering, dancing, and revelry on a flat piece of ground on which the tents of the two parties are ranged opposite each other between two slight eminences, about a bow-shot apart. The scene is looked forward to by every one with pleasant anticipations, and is spoken of as one of such great excitement that they hardly sleep during the time it lasts.

About the 20th of July this friendly meeting is dissolved, the Nu-na-tang'-meun ascending the Colville homewards, and the others descending its eastern mouth to pursue their journey to O-lik'-to, Point Berens. In consequence of their occupying a great deal of time in hunting to provide supplies for the remainder of the journey, they spend four or five days in this short distance, which does not exceed twenty miles. Proceeding from Point Berens they travel four sleeps, as marked in red ink on the chart, to a place called Ting-o-wai'-ak (Boulder Island of Franklin), where the tents are pitched and the women and children left. Three boats are then selected, and additional benches placed in each for the accommodation of its crew, now increased to fifteen, including one or two women. The fifth sleep is within a short distance of Barter Point, from which they start prepared for a hostile or a friendly meeting, as the case may be, but it is uniformly the latter, at least of late years. The conduct of the Point Barrow people in their intercourse with those of the Mackenzie, or rather Demarcation Point, seems to be very wary, as if they constantly kept in mind that they were the weaker party, and in the country of strangers. They describe themselves as taking up a position opposite the place of barter on a small island to which they can retreat on any alarm, and cautiously advance from it making signs of friendship. They

say that great distrust was formerly manifested on both sides by
the way in which goods were snatched and concealed when a
bargain was made; but in later years more women go, and they
have dancing and amusements, though they never remain long
enough to sleep there. They state that on leaving Barter Point
the wind is always easterly, and making sail on their boats, they
can go to sleep. On the first day they pick up the women and
children with their tents, and return to Point Berens on the second.
They now cross Harrison Bay in a direct line before the breeze to
Cape Halkett, about the 10th of August, some taking the route through
the rivers by which they had gone eastward, and others proceeding
along the sea coast. Should the previous whaling season have been
successful, they spend the time until September in fishing and
catching deer ; but should the opposite have been the case, they
make no delay beyond what is necessary for procuring supplies to
bring them back to Nu-wúk, in order to make up in the autumn
for the deficiency of the summer.

The traffic, which is the main object of this yearly journey, has
been already alluded to, but some more details of it may not prove
uninteresting. At the Colville, the Nu-na-tang'-meun offered the
goods procured at So-su'-a-ling on Kotzebue Sound from the
Asiatics, Kokh-lit' en'-yu-in, in the previous summer, consisting of
iron and copper kettles, women's knives (o-lu'), double-edged
knives (pan'-na), tobacco, beads, and tin for making pipes; and
from their own countrymen on the Ko'-wak River, stones for
making labrets, and whetstones, or these ready made, arrow-heads,
and plumbago. Besides these are enumerated deer and fawn-skins,
and coats made of them, the skin, teeth, and horns of the im'-na
(argali ?), black fox, marten, and ermine-skins, and feathers for
arrows and head-dresses. In exchange for these, the Point Barrow
people (Nu-wung'-meun) give the goods procured to the eastward
the year before, and their own sea-produce, namely, whale or seal-
oil, whalebone, walrus-tusks, stout thong made from walrus-hide,
seal-skins, &c., and proceed with their new stock to Point Barter.
Here they offer it to the Kan'g-ma-li en'-yu-in, who may be
called for distinction Western Mackenzie Eskimo, and receive in
return, wolverine, wolf, imna, and narwhal skins (Kil-lel'-lu-a),
thong of deer-skin, oil-burners, English knives, small white beads,
and latterly guns and ammunition. In the course of the winter
occasional trade takes place in these with the people of Point Hope,
but most of the knives, beads, oil-burners, and wolverine-skins, are
taken to the Colville the following year, and, in the next after,
make their appearance at Kotzebue Sound and on the coast of Asia.

From what we know positively of the trade thus far, we are inclined to believe there is a tolerably regular yearly communication between each Eskimo tribe and their neighbours of the same race on either side. It seems highly probable the pan'-na, or double-edged knife, described by Sir W. E. Parry as in use among the tribe he met at Winter Island, may have been of Siberian origin, from being of the same form and identical in name with that brought by the Asiatics to Hotham Inlet, where they receive in return oil-burners, or stone lamps, which we have often seen in their tents in 1848-9, of a shape corresponding exactly with the drawing in that gentleman's journal of his second voyage; they bear also a similar name, kŏd'-lan, and are said to be brought from a very distant eastern country. Supposing a knife of this kind made in Siberia, to be carried at the usual rate, we compute it would not arrive at Winter Island before the sixth year, and, having been exchanged the year before for a stone lamp, this might come into the hands of the Asiatics on the ninth. The knife would remain the first winter in the possession of the Reindeer Tuski (or Tsau'-chu), the second with the inland Eskimo, Nu-na-tang'-meun, the third at Demarcation Point with the Kang'-ma-li-meun, the fourth with the East Mackenzie or the Cape Bathurst tribes, and on the fifth possibly fall into the hands of the people who make the lamps. The lamp, returning the same way, would remain the sixth winter at Cape Bathurst, the seventh at Demarcation Point, the eighth at Point Barrow, the ninth in the interior, and be received by the Asiatics on the following summer.

For a very large portion of our information, we have been indebted to a man called Erk-sin'-ra, who has sustained a most excellent character throughout the whole time the *Plover* remained at Point Barrow. He drew the coast-line eastward as far as he knew it, giving the names of many places, some of which he described so minutely as to be undeniably identified with those mentioned in Sir J. Franklin's journal, and laid down in his chart. Erk-sin'-ra's coast-line has been drawn in red, parallel to that copied from the Admiralty chart, and a dotted line marks each place where the two were made out clearly to correspond. What seemed to us most singular was, that whilst his description of the coast agreed so minutely in many particulars with the narrative and chart of Messrs. Dease and Simpson, he denied the existence of the Pelly Mountains, and maintained most positively that there are no hills on the west side of the Colville visible from the sea; and at length said, "We never saw them, but perhaps you might with your long spy-glasses." He was the head man of the first party Commander

Pullen met at Point Berens on the 11th of August, 1849, and gave O'-lik-to as the name of the place where the post was erected. By a letter dated H.M.S. *Investigator*, 8th of August, 1850, received from a native of Point Barrow, to whom it had been given at Point Drew, that ship must have passed Point Berens on the 9th or 10th of August, when she also was seen by Erk-sin'-ra. As he was on both these occasions on his return from that bartering-place, the first week in August may be confidently assumed as the usual time of the two tribes meeting at Barter Point.

Among the few remarkable features of this dreary coast is a large stone, about four sleeps from Point Barrow, near Point Tangent, giving the name of Black Rock Point to the projecting land off which it lies. It is mentioned by Mr. J. Simpson as the only stone of large size he met with on this part of his journey. The natives assert it is a "fire stone," and fell from the sky within the memory of people now living. No one saw it fall; but one woman, about sixty years of age, said she travelled that way yearly as a girl, when there was no stone there, and that in returning one summer, her people were much surprised to see it, and believed it had fallen from the sky. Should it prove a meteoric stone, the story of its age might be true enough; but at present it is doubtful. It is said to enlarge and present a full rounded appearance at times, when deer are plentiful in the neighbourhood, as it feeds upon them, killing and devouring a great many at a time. No doubt those animals are instinctively guided in their migrations by particular states of the atmosphere; and as the tides are much influenced by the winds, it is not impossible that they should most abound in that locality when the tide is low, giving an apparent increase to the size of the stone.

We were anxious to get the history of the "Old Huts," marked by Sir J. Franklin in longitude 146° 20' w., but could ascertain distinctly no more than that they were the remains of an ancient Kang'-ma-li settlement. In connection with this, our informants gave an account of the modern origin of the trade at Barter Point, agreeing with that given by Sir J. Franklin, to the effect that it was established within the memory of people recently dead, whilst their intercourse with the inland people by the Colville is of ancient date. But from their having traditions of the Eastern people relating to a remote period, we think it probable that it was only renewed in recent times, having been previously kept up by a tribe inhabiting the "Old Huts," whose parties visited the Colville on the west, and met the Mackenzie people on the east of their own country. From the well-known hostility of the Red Indians to the

Eskimos, it may be conjectured that the settlement was destroyed by them and the inhabitants put to death; and that after some time had elapsed, the people of Point Barrow would be induced to extend their journeys eastward farther in search of those whose goods they had been accustomed to receive, and at length meeting with other people, none of whom they had ever before seen, the establishment of a regular trade, as at present existing at Barter Point, would be the result.

Point Hope is generally visited by parties in the winter, who perform the journey in fifteen to twenty days, returning to Nu-wŭk at the end of two moons. From that Cape, therefore, to a little beyond Barter Point, a distance of about 600 miles, is the extent of coast with which the Point Barrow people are actually acquainted, and their personal knowledge of the interior may be said to extend to fifty miles. But besides this they also know, by report, the names of more distant countries and their inhabitants; thus the people they trade with at Barter Point are called Ka'ng-ma-li en'-gu-in, whose winter huts are probably at Demarcation Point; among them they have occasionally seen a few Ko-pan'g-meun, Great River (Mackenzie) people, whom they distinguish by having a tattooed band across the face. Beyond the Mackenzie is a country called Kit-te-ga'-ru, and farther still, but very distant, one inhabited by the people who make the stone lamps before spoken of. So far they speak with confidence; and then relate the story of a singular race of men living somewhere in that direction, who have two faces, one in front and the other at the back of the head. In each face is one large eye in the centre of the forehead, and a large mouth armed with formidable teeth. Their dogs, which are their constant companions, are similarly provided with a single eye in each. This fable seems to refer to the tribe of Indians who are said by their neighbours to see the arrows of their enemies behind them.

Of the Indians they know but little personally, having only seen a few on rare occasions; but they appear to know them well· by report, both from the Ka'ng-ma-li-meun and Nu-na-tan'g-meun. Under the general term It'-ka-lyi, they describe them as a dangerous people, well armed with guns, who reside in the mountainous districts far away to the south and east of the Colville. The inland Eskimo also call them Ko'-yu-kan, and divide them into three sections or tribes, two of which they know, and say they have different modes of dancing. One is called It'-ka-lyi, and inhabits the It'-ka-ling River, east of the Colville; the second, It-kal-ya'-ru-in, whose country is farther south; and the third, whom they have never seen, but only heard of as the people who

barter wolverine-skins, knives, guns, and ammunition to the
Eskimo at Herschel Island, for Russian kettles, beads, &c.,
together with whalebone and other sea-produce. These three
tribes, they further say, are all dressed alike, and are fierce and
warlike, but not cannibals like other Indians they have heard of.
They are, without doubt, the mountain Indians to whom Sir J.
Franklin makes frequent allusion in his narrative of his journey
westward from the Mackenzie River, a tribe who have had but
little intercourse with the Hudson's Bay Company; and Mr. J.
Simpson, travelling the same coast in 1837, also mentions them as
but little known. As the name Ko'-yu-kan, by which they are
known at Point Barrow, is the same as that given to the tribe in
whose treacherous attack on the Russian post at Darabin Lieutenant Barnard lost his life in 1851, and as some of their coasts and
other portions of dress offered for sale at the *Plover*, in 1852, were
of the same make and material as the suit in the possession of Mr.
Edward Adams of the *Enterprise*, the companion of Lieutenant
Barnard, there can be little doubt they are one and the same
people. If, as seems probable, they are also the same who destroyed
the Hudson's Bay post in 1839, in latitude 58°, they occupy a great
extent of country between the Colville and Mackenzie Rivers, and
range from near Sitka to the Arctic Sea. It is at all times desirable
that great caution should be used in drawing inferences from mere
sounds in an unwritten language which is but partially known, yet
it seems worthy of remark, that the Eskimo word Kök, a river,
if prefixed to the name Yu-kon, will bear a strong resemblance to
the name Ko'-yu-kan, given by them to the Indians inhabiting the
country through which the You-kon flows. They also know by
report the people of Cape Prince of Wales, Kin'g-a-meun, and the
Kokh-lit' en'-yu-in, Asiatics, who come to Kotzebue Sound yearly.

Some traditions they have besides which refer to a land named
Ig'-lu, far away to the north or north-east of Point Barrow. The
story is, that several men, who were carried away in the olden
time by the ice breaking under the influence of a southerly wind,
after many sleeps arrived at a hilly country inhabited by a people
like themselves who spoke the same language. They were well
received and had whales'-flesh given them to eat. Some of these
wanderers found their way back to Point Barrow, and told the tale
of their adventures. After some time, during a spring when there
was no movement in the sea-ice, three men set out to visit this
unknown country, taking provisions on their backs; and having
performed their journey without mishap, brought home confirmation of the previous accounts. Nothing further could be learned

concerning this northern expedition except that each man wore out three pair of mocassin soles in the journey ; and since then there has been no communication with the Ig'-lun Nu'-na, but they believe some others who have been carried away on the ice may have reached it in safety.

We could never find any who remembered having seen Europeans before Mr. J. Simpson's visit in 1837, but had heard of them as Ka-blu'-nan from their eastern friends ; more recently they heard a good deal of them from the inland tribes as Tan-ning or Tan'-gin. This probably refers to the Russians, who have regular bath days at their posts, and is derived from tan-ni'kh-lu-go, to wash or cleanse the person. They also apply other names to us, apparently of their own invention ; one is E-ma'kh-lin, sea men (this is the name of the largest of the Diomede Islands) ; another is Sha-ke-na-ta'-na-meun, people from beneath the sun (en'-gu-in a-ta'-ne Sha-ke'-nik) ; but the most common one is Nel-lu-an'g-meun, unknown people (nel-lu-a'-ga, I do not know).

To themselves they apply the word En-yu-in, people, the plural of ĕ-nyu'k, a person of any nation, prefixing, when necessary, the name of their nu-na or country, as, Nu-wu'ng-meun, that is, Nu-wu'k En'-yu-in, Noo-wook or Point Barrow people ; Ing-ga-lan'-da-meun, Englishmen. Lately those met with in Grantley Harbour and Port Clarence have adopted the epithet Es-ki-mo'.

In addition to the notice of the phases of the moon, they possess sufficient knowledge of the stars to point out their position in the heavens at particular seasons, and we believe use them as guides sometimes in travelling. They look upon them as fiery bodies, as proved in their estimation by the shooting stars, which they look upon as portions thrown off by the fixed ones. They form them into groups, and give them names, many of which they explain. The star Aldebaran, with the cluster of the Hyades, and other smaller ones around, are called Pa-chúkh-lu-rin, " the sharing-out " of food, the chief star representing a polar bear just killed, and the others the hunters around, preparing to cut up their prize, and give each hunter his portion. The three stars in Orion's Belt are three men who were carried away on the ice to the southward in the dark winter. They were for a long time covered with snow, but at length perceiving an opening above them, they ascended farther and farther until they became fixed among the stars. Another group is called the " house building," and represents a few people engaged in constructing an ig-lu, or winter hut. But perhaps their most complete myth refers to the sun and moon, who, they say, are sister and brother. Given as we received it,

it runs as follows:—" A long time ago, in a country far away to
the eastward, called Pin'g-ŏ, the people held a winter festival, when
one of the women, tired of dancing, left the company and retired to
rest in her own hut. Before she had gone to sleep, she perceived
some one enter, who blew out the light, and lay down beside her.
Being desirous to know who her stealthy visitor was, she smeared
her hands with soot from the lamp within her reach, and secretly
blackened his body, that she might know him again among the
dancers. After he had gone, she returned to the dance-house, and
peeping in, saw to her horror that the man whose person she had
marked was her own brother. She retired in great grief to the
open air ; but soon returning to the dance-house, she went into the
middle of the assembly, and with a woman's knife (o-lú) cut off her
left breast, which she gave her brother, saying, 'All this it is good
that you should eat.'¹ They then went out, and both ascended
slowly towards the heavens in a circular path, he with his dog going
first and she following, and when nearly out of sight separated, the
man, by name Nel-lu-kat'-si-a Tád-kak, to become the moon, and his
sister, Sigh-rá-a-na, to become the sun, still dripping with her own
gore, as may be seen occasionally in cloudy weather, when she looks
red and angry." The moon is considered cold and covered with
snow, on the white surface of which may be traced at the full the
figure of the man perpetually travelling with his dog, whilst the lady
sun enjoys the warmth of an eternal summer."

    In some of their pursuits necessity compels the men of different
establishments to combine their strength, as in taking the whale,
and in such circumstances, some must take the lead. It would
seem an easy step from this to the permanent ascendency of individuals over the others, and some have accordingly considerable
weight in the community ; but there is nothing among them
resembling acknowledged authority or chieftainship. A man who
has a boat out in the whaling season, engages a crew for the time ;
but while in the boat he does not appear to have any control over
them, and asks their opinion as to where they should direct their
course, which, however, they generally leave him to determine, as
well as to keep the principal look-out for whales. The chief men
are called Ome'liks (wealthy), and have acquired their position by
being more thrifty and intelligent, better traders, and usually better
hunters, as well as physically stronger and more daring. At the
winter and summer festivals, when the people draw together for

---

¹ This is not given as a literal translation, but we believe it conveys the
meaning. The Eskimo words are " ta-man'g-ma mam-mang-mang-an'g-ma
nigh'-e-ro."

enjoyments, proficiency in music, with general knowledge of the customs and superstitions of their tribe, give to the most intelligent a further ascendancy over the multitude; and this sort of ascendancy once established, is retained without much effort. As they combine to form a boat's crew to pursue a common prey, so will they unite to repel a common enemy, but it is only when danger is common they will so unite: their habits of life leaving them perfectly free from the control of others, and making them dependent solely on their own individual exertions for a livelihood; they are bound together as a society only by ties of relationship and a few superstitious observances, and have no laws or rules excepting what custom has established in reference to the spoils of the chase. It cannot be doubted that their Ome'liks have considerable influence, more especially over their numerous relations and family connections, and may use some art to maintain and extend it; yet O-mig'-a-loon, the most influential man at Nu-wu'k, the same who headed the party against Commander Pullen at Point Berens, after informing us that a lad of eighteen had deceived us, and got food by telling a false tale of distress, would not for some time repeat his statement in the presence of the youth.

Invisible spirits (*sing.* turn'-gak; *plural*, turn'-gain) people the earth, the air, and the sea; and to them they apply similar notions of equality, attributing to none superior power, nor have they even a special name for any that we could learn. These turn'-gain are very numerous, some good, some bad; they are sometimes seen, and then usually resemble the upper half of a man, but are likewise of every conceivable form. Their belief in ghosts seemed proved by the circumstance that two young girls who left the ship in the twilight of a short winter's day, turned back in breathless haste on seeing a sledge set up on end near the path to the village. They told the story of themselves next day, saying they were frightened, having mistaken the sledge, which was not there in the daytime when they had passed, for a turn'-gak. They are concerned in the production of all the evils of life, and whatever seems inexplicable is said to be caused by one of them. One causes a bad wind to blow, so that the ice becomes unsafe; another packs the ice so close on the surface of the sea, that the whales are smothered; and a third strikes a man dead in the open air, without leaving any mark on his body; or a fourth draws him by the feet into the bowels of the earth. These are evil genii; and the good ones are little better, as they are very liable to get offended and turn their backs on suffering humanity, leaving it at the mercy of the worse disposed. Their dances and ceremonies are all intended to please, to cajole, or to frighten these spirits. The most curious ceremony that came under observation was performed at the village in the course of the last winter, when food had become very scarce in consequence of the ice continuing very close from a long continuance of north-westerly winds. On the sea beach, close to one of the dance-houses, a small space was cleared, and a fire of wood made, round which the men formed a ring and chanted for some time, without dancing or the usual accompaniment of the tambourine. One of the old men then stepped towards the fire, and in a coaxing voice tried to persuade the evil genius, from whose baleful influence the people were suffering, to come under the fire to warm himself. When he was supposed to have arrived, a vessel of water, to which each man present had contributed, was thrown upon the fire by the old man, and immediately a number of arrows sped from the bows of the others into the earth where the fire had been, in the full belief that no turn'-gak would stop at a place where he received such bad treatment, but would soon depart to some other region, from which, on being detected, he would be driven away in a similar manner. To render the effect still greater, three guns were fired in different directions, to alarm the spirits of the air, and make them change the wind. For the same object they several times requested the ship's guns, eighteen-pounders, to be fired against the wind.

When our poor friend O-mis-yu-a'-a-run, commonly called the water-chief, from having accused us of stealing the water from the village, was carried away with two others on the ice to near Cape Lisburne, in the beginning of the winter, his wife had a thin thong of seal-skin stretched in four or five turns round the walls of the ig-lu, and anxiously watched it night and day until she heard of her husband's fate. They believe that so long as the person watched for is alive and moves about, his turn'-gak causes the cord to vibrate, and when at length it hangs slack and vibrates no longer, he is supposed to be dead. Having heard something of the hourly observations of the movements of a magnet suspended by a thread in the observatory, the old dame sent Erk-sin'-ra to see if its movements had any connection with her husband's case.

Thunder is a rare occurrence at Point Barrow, but not altogether unknown to its inhabitants, and they say the sound of it is caused by a man spirit, who dwells with his family in a tent far away to the north. This Eskimo representative of Jupiter Tonans is an ill-natured fellow who sleeps most of his time; and when he wakes up he calls to his children to go out and make thunder and lightning by shaking inflated seal-skins and waving torches, which they do with great glee until he goes to sleep again.

They do not entertain any clear idea of a future state of existence, nor can they apparently imagine that a person altogether dies. Although death is a subject they dislike to talk of, we have heard the sentiments of several upon this, and the nature of the soul. About the last they differ a good deal, but they all agree in looking upon death as the greatest of human evils, and would invariably " rather bear the ills they have, than fly to others that they know not of." The soul is a turn'-guk, they say, seated in the breast, or rather in the lungs, and seems closely allied to the breath ; from it emanate all thoughts, which as they rise the tongue gives utterance to. Even as to its unity they hold different notions, for one person told us a man had four turn'-gain in his breast; and another, that wherever a man went there was in the ground beneath him his " familiar spirit," which moved as he moved, and was only severed from him in death. However this may be, in death the body sleeps and the spirit descends into the earth to associate with those which have gone before, and subsists on bad food, such as roots, stones, and mosquitoes.

In order not to offend the spirits of the departed, their bodies are wrapped in skins and laid on the earth beside others already there, with the head to the east at Point Barrow ; but for this direction there is no general rule. As his clothes and other portions of property he habitually used, including the sledge on which he was carried, would bring ill-luck to any one else who took them, they are left with the body in a torn or broken state, and the family to which he belonged keep within the hut for five days, not daring to work lest the spirits should be offended ; and instances can be readily adduced where they believe death to have happened to persons who infringed the custom of mourning five days. Diseases are also considered to be turn'-gaks ; and so hurtful do they think the touch of a corpse, that it is unwholesome to smoke from the same pipe or drink out of the same cup with any one who was the wife, mother, or other near relative of a deceased person ; this, they say, is because these relatives from tending the sick person become tainted by his breath, and another by using the same pipe or cup might acquire the disease.

JOHN SIMPSON, Surgeon, R.N.

u

REPORT OF THE ANTHROPOLOGICAL INSTITUTE.

To the Council of the Anthropological Institute of Great Britain and Ireland.

YOUR Committee, to whom was referred the annexed letter from the Royal Geographical Society, have agreed to the following Report :—

24th May, 1872.

SIR,—The President and Council of the Royal Geographical Society, after a careful consideration of a Report drawn up by a Committee of Arctic Officers[1] belonging to their body, having come to the conclusion that the time has arrived for once more representing the important results to be derived from Arctic exploration to Her Majesty's Government ; I have been directed to request that the following remarks may be laid before the President and Council of the Anthropological Institute.

In a letter to me signed by the late Mr. George E. Roberts, and dated May 8th, 1865, he was instructed to say that the Council of the Anthropological Society viewed with the deepest interest the prospect of an Arctic exploring expedition ; believing that great advantage to their science would ensue from such an undertaking.

Strengthened by the willingness expressed by the Council of the Anthropological Institute to co-operate with the Royal Geographical Society in adopting such measures as might be considered advisable to induce Her Majesty's Government to accede to the proposal of fitting out an Arctic expedition, and by other expressions of cordial approval received from kindred scientific Societies, Sir Roderick Murchison brought the subject of North Polar exploration to the notice of the Duke of Somerset, then First Lord of the Admiralty, in a letter dated 19th of May, 1865 ; and the subject was discussed between his Grace and a deputation from the Council of the Royal

---

[1] Sir George Back, Admiral Collinson, Admiral Ommanney, Admiral Sir L. McClintock, Admiral Richards, Captain Sherard Osborn, Mr. A. G. Findlay, Mr. Clements Markham (Sec.).

Geographical Society, in an interview which took place on the 20th of June in the same year.

But at that time there was some difference of opinion among Arctic authorities on the subject of the best route to be adopted, and the Duke said that he would wish to be in possession of the results of the Swedish Expedition then engaged in exploring Spitzbergen, and of other information, before he could recommend an Arctic exploring expedition to the consideration of the Government.

In consequence of the view taken by his Grace, the Council of the Royal Geographical Society have carefully watched the results of expeditions undertaken by foreign countries, in order to be in a position to recommend one route as undoubtedly the best, before again pressing the subject upon the attention of the Government. Seven years have now passed, and during that time additional experience has been accumulated by the Swedes and Germans, which has enabled the Council to form an opinion that justifies a renewal of their representation made in 1865. The distinguished Arctic officers who are Members of the Geographical Council, and who have carefully considered the evidence accumulated since 1865 in a special Committee, are now unanimously of opinion that the route by Smith Sound is the one which should be adopted with a view to exploring the greatest extent of coast-line, and of securing the most valuable scientific results. The conclusion thus arrived at by authorities of such eminence has placed the Royal Geographical Society in a position which will enable its Council to represent to the Government that the conditions are now fulfilled which the First Lord of the Admiralty deemed essential in 1865, before he could entertain the project of North Polar Exploration.

I am, therefore, instructed to represent the very great importance of stating the scientific results to be derived from the exploration of the unknown North Polar Region in full detail, even in a first preliminary communication to the Government. It is believed that the success of any representation will depend to a considerable extent on the force and authority with which that portion of it is prepared, which enumerates the scientific results to be derived from the proposed expedition. I am to request that you will submit these views to the President and Council of the Anthropological Institute, and that they will be so good as to cause a statement to be drawn up and furnished to the Council of the Royal Geographical Society, embodying their views, in detail, of the various ways in which the Science of Anthropology would be advanced by Arctic exploration.

I enclose, for the information of the President and Council of the

U 2

Institute, copies of a Memorandum which has been prepared upon the subject, and of the papers which were read by Captain Sherard Osborn in 1865 and 1872, advocating a renewal of Arctic exploration.

I have the honour to be, Sir, your obedient servant,

CLEMENTS R. MARKHAM.

To the Secretary of the Anthropological Institute.

REPORT *of the* ARCTIC COMMITTEE[1] *of the* ANTHROPOLOGICAL INSTITUTE.

THE knowledge already acquired of the Arctic Regions, leads to the conclusion that the discovery of the unknown portion of the Greenland coasts will yield very important results in the science of Anthropology. Although barely one-half of the Arctic Regions has been explored, yet abundant traces of former inhabitants are found throughout their most desert wastes, where now there is absolute solitude. These wilds have not been inhabited for centuries, yet they are covered with traces of wanderers or of sojourners of a bygone age. Here and there, in Greenland, in Boothia, on the shores of America, where existence is possible, the descendants of former wanderers are still to be found. The migrations of these people, the scanty notices of their origin and movements that are scattered through history, and the requirements of their existence, are all so many clues which, when carefully gathered together, throw light upon a most interesting subject. The migrations of man within the Arctic zone give rise to questions which are closely connected with the geography of the undiscovered portions of the Arctic Regions.

The extreme points which exploration has yet reached on the shore of Greenland, are in about 80° on the west, and in 76° on the eastern side; and these two points are about 600 miles apart. As there are inhabitants at both these points, and they are separated by an uninhabitable interval from the settlements further south, it may be inferred that the unknown interval further north is or has been inhabited. On the western side of Greenland it was discovered, in 1818, that a small tribe inhabited the rugged coast, between 76° and 79° N.; their range being bounded on the south by the glaciers of Melville Bay, which bar all progress in that direction; and on the north by the Humboldt glacier; while the

---

[1] This Committee consisted of Sir John Lubbock (President), Professor Busk, Captain Sherard Osborn, Captain Bedford Pim, Col. Lane Fox, Mr. Clements Markham, Mr. Flower, and Mr. Brabrook.

Sernik-soak, or great glacier of the interior, confines them to the sea-coast. These " Arctic Highlanders " number about 140 souls, and their existence depends on open pools and lanes of water throughout the winter, which attract animal life. Hence, it is certain that where such conditions exist man may be found. The question whether the unexplored coast of Greenland is inhabited, therefore, depends upon the existence of currents and other conditions such as prevail in the northern part of Baffin's Bay. But this question is not even now left entirely to conjecture. It is true that the " Arctic Highlanders," told Dr. Kane that they knew of no inhabitants beyond the Humboldt glacier, and this is the furthest point which was indicated by Kalli-hirua (the native lad who was on board the *Assistance*) on his wonderfully accurate chart. But neither did the Eskimo of Upernivik know anything of natives north of Melville Bay until the first voyage of Sir John Ross. Yet now we know that there either are or have been inhabitants north of the Humboldt glacier, on the extreme verge of the unknown region; for Morton (Dr. Kane's steward) found the runner of a sledge made of bone lying on the beach on the northern side of it. There is a tradition, too, among the " Arctic Highlanders," that there are herds of musk-oxen far to the north, on an island in an iceless sea. On the eastern side of Greenland there are similar indications. In 1823, Captain Clavering found twelve natives at Cape Borlase Warren in 76° N.; but when Captain Koldewey wintered in the same neighbourhood in 1869 none were to be found, though there were abundant traces of them and ample means of subsistence. As the Melville Bay glaciers form an impassable barrier, preventing the " Arctic Highlanders " from wandering southwards on the west side; so the ice-bound coast on the east side, between Scoresby's discoveries and the Danebrog Isles, would prevent the people seen by Clavering from taking a southern course. The alternative is that, as they were gone at the time of Koldewey's visit, they must have gone north.

These considerations lead to the conclusion that there are or have been inhabitants in the unexplored region to the north of the known parts of Greenland. If this be the case, the study of all the characteristics of a people who have lived for generations in a state of complete isolation, would be an investigation of the highest scientific interest.

Light may not improbably be thrown upon the mysterious wanderings of these northern tribes, traces of which are found in every bay and on every cape in the cheerless Parry group; and these wanderings may be found to be the most distant waves of

storms raised in far off centres, and among other races. Many
circumstances connected with the still unknown northern tribes
may tend to elucidate such inquiries. Thus, if they use the *iglu*
they may be supposed to be kindred of the Greenlanders; snow-
huts will point to some devious wanderings from Boothian or
American shores; while stone *yourts* would indicate a march from
the coast of Siberia, across a wholly unknown region. The method
of constructing sledges would be another indication of origin, as
would also be the weapons, clothes, and utensils. The study of the
language of a long isolated tribe will also tend to elucidate questions
of considerable interest; and its points of coincidence and diver-
gence, when compared with Greenland, Labrador, Boothian, and
Siberian dialects, will lead to discoveries which, probably, could not
otherwise be made. Dr. Hooker has pointed out that the problem
connected with the Arctic flora can probably be solved only by a
study of the physical conditions of much higher latitudes than have
hitherto been explored. In like manner, the unsolved puzzles connected with the wanderings of man within the Arctic zone may
depend for their explanation upon the clues to be found in the
conditions of a tribe or tribes in the far north.

These are speculations which the results gained by Polar discovery
would probably, but not certainly, show to be well founded. But
there are other investigations which would undoubtedly yield valuable materials for the student of man. Such would be carefully
prepared notes on the skulls, the features, the stature, the dimensions of limbs, the intellectual and moral state of individuals
belonging to a hitherto isolated and unknown tribe; also on their
religious ideas, on their superstitions, laws, language, songs, and
traditions; on their weapons and methods of hunting; and on their
skill in delineating the topography of the region within the range
of their wanderings. There are also several questions which need
investigation, having reference to marks and notches upon arrows
and other weapons, and to their signification. A series of questions
has been prepared by Dr. Barnard Davis, Mr. Tylor, Col. Lane Fox,
and others, on these and other points,[1] attention to which would

1.   1. Instructions of Dr. Barnard Davis.
    2. Enquiries as to Religion, Mythology, and Sociology of Eskimo Tribes, by
E. B. Tylor, Esq., F.R.S.
    3. Enquiries relating to Mammalia, Vegetation, &c., by W. Boyd Dawkins,
Esq., F.R.S.
    4. Enquiries into Customs relating to War, by Col. A. Lane Fox.
    4a. Enquiries relating to certain Arrow-marks and other Signs in use among
the Eskimos.
    4b. Enquiries relating to Drawing, Carving, &c., by Col. A. Lane Fox.
    5. Enquiries as to Ethnology, by A. W. Franks, Esq.
                                                        6. Enquiries

undoubtedly result in the collection of much exceedingly valuable information.

The condition of an isolated tribe, deprived of the use of wood or metals, and dependent entirely upon bone and stone for the construction of all implements and utensils, is also a subject of study with reference to the condition of mankind in the stone age of the world; and a careful comparison of the former, as reported by explorers, with the latter, as deduced from the contents of tumuli and caves, will probably be of great importance in the advancement of the science of man.

For the above reasons there cannot be a doubt that the despatch of an expedition to discover the northern shores of Greenland would lead to the collection of many important facts, and to the elucidation of deeply interesting questions connected with anthropology.

APPENDIX.

QUESTIONS FOR EXPLORERS.

(With Special Reference to Arctic Exploration.)

1. GENERAL. By J. BARNARD DAVIS, M.D., F.R.S.

1. *Names of Tribes*, indicating their divisions, and at the same time marking any peculiarities of any kind which distinguish them. This will embrace Tribal marks.

2. *Stature of Men and Women.*—For this purpose the traveller should be provided with a measuring-tape or other instrument. Measure twenty-five of each, if he can.

3. *Colours of Skin, Eyes, and Hair.*—These are easily determined by Broca's Tables.

4. *Hair, Texture of and Mode of Wearing.*—Specimen locks, tied up separately and accurately labelled, if possible.

5. *Deformations* carefully observed and accurately described. Those of the heads of infants impressed in nursing, if any; those of the teeth produced by chipping, filing, &c.; those of the skin done by tattooing, incisions, scars, wheals, &c., correctly described.

6. *Crania* diligently collected. These should always be procured as perfect as possible, never leaving anything behind, particularly

---

6. Enquiries relating to the Physical Characteristics of the Eskimo, by Dr. J. Beddoe.
7. Further Ethnological Enquiries, by Professor W. Turner.
8. Instructions suggested by Captain Bedford Pim, R.N.

not lower jaws and teeth. On collection, they should be at once marked with tribal name, *in ink* if possible, to prevent confusion.

7. *Diseases.*—Careful observations upon their names, natures, peculiarities, &c., and their modes of treatment, if they can be ascertained.

8. *Careful Observations* of the habits and modes of life of the people : their social, intellectual, and moral state.

9. *Portraits*, by drawing or photography, should not on any account be omitted, if attainable.

10. *Articles* of dress, implements. &c., should be collected.

11. *Systems of Relationship.*—(See ' Journal of Anthropological Institute ' (vol. i. p. 1), paper by Sir J. Lubbock, President.)

12. *Language.*—As complete a vocabulary as circumstances will allow should be recorded.

2.—ENQUIRIES *as to* RELIGION, MYTHOLOGY, *and* SOCIOLOGY *of* ESKIMO TRIBES. By E. B. TYLOR, Esq., F.R.S.

1. What ideas have they as to souls and other spirits? What do they think of dreams and visions? are they appearances of spirits? Are trances, &c., set down to exit of soul? Are hysterics, convulsions, &c., ascribed to demoniacal possession?

2. Does the soul continue to exist after death? is there any difference made in the fate of souls? and, if so, is the difference due to their conduct in life? Is there any transmigration of souls?

3. Are there spirits in rocks, springs, mountains, &c.? if so, what are their appearance, functions, and names?

4. Are there any great gods believed in (*e.g.*, a sun god), &c.? Especially is there one called Torngarsuk, or Great Spirit?

5. What prayers, sacrifices, fasts, ceremonial dances, religious festivals, &c., have they?

6. What sorcerers or seers have they? how brought up, and practising what crafts? What necromancy, divinations, and other magic arts have they?

7. What legends of gods and heroes have they? What stories which seem to relate to personified natural phenomena, sun, moon, &c.?

8. What actions and dispositions are considered good and bad, virtuous and vicious? Does public opinion make much difference in treatment of virtuous and vicious? Are there any set laws and penalties? what restraint is there on theft, murder, adultery, &c.? Do acts count as criminal differently when done on a member of the tribe or foreigners? What is the native law or custom as to

vengeance? What are the laws or customs as to marriage, inheritance, and clanship?

9. What recognition of chiefship and what form of civil government can be traced? Are the old men rulers, and do the strong men displace them? What is the treatment of women and children, and of the sick and aged?

3.—QUESTIONS *relating to the* MAMMALIA, *the* VEGETATION, *and the* REMAINS *of* ANCIENT RACES. By W. BOYD DAWKINS, M.A., F.R.S.

Where do the Eskimos obtain the ivory which they use for handles to their scrapers and for other purposes? Besides the walrus ivory they use the tusks of the mammoth: how do they know where to seek for these, and have they any legends in connection with them? The conditions under which these tusks occur in the regions bordering on the great Arctic Sea are of the highest importance as throwing light on similar remains in Northern and Central Europe. The bones and teeth of the smaller animals, which most probably occur in the same strata as the mammoth ivory, should be preserved, for there is reason to believe that at a time comparatively recent, zoologically speaking, the climate of the extreme north was far less severe than now.

The sources from which the Eskimos obtained their wood should be carefully ascertained. Is it drift-wood brought down by great rivers, like the Obi or the Mackenzie, from more southern latitudes? or is it derived from ancient forests which once flourished where at the present time no trees will grow?

Have the Eskimos any legends relating to other lands than those in which they now live; in other words, what was their golden age?

Have the Eskimos any legends relating to the musk-sheep, *Ovibos moschatus?*

4.—ENQUIRIES *into* CUSTOMS RELATING *to* WAR. By Col. LANE FOX.

1. *Tactics*—Have the tribes any disposition or order of battle? are the young or the weak placed in front? are they courageous? have they any war cries, war songs, or war dances, and if so give a detailed account of them? Do they employ noise as a means of encouragement, or do they preserve silence in conflict? Do they stand and abuse each other before fighting, or boast of their warlike achievements? Do they rely on the use of missile-weapons or hand-weapons? have they any special disposition for these in battle? have they any knowledge of the advantages of ground or position in battle, as sug-

gested by Capt. Beechey? have they any sham fights with blunt and pointless weapons, such as are described by Vancouver in Hawaii and amongst the Hottentots? How is the march of a party conducted? do they move in a body with a broad front or in file, and do they send forward advanced parties? do they make night attacks? have they any stratagems for concealing their trail from the enemy? Have they any superstitious customs or omens in connection with war, and if so give an account of them? What is the meaning of the custom of shooting an arrow with a tuft of feathers attached, mentioned by Capt. Beechey, and supposed to be a declaration of war? (the custom of shooting an arrow towards an enemy as a declaration of war formerly existed in Persia.) Do they employ treachery, concealment, or ambush, and if so, what is their usual mode of proceeding? Are their dogs employed in war? Are their treaties with other tribes binding? Do they form alliances with other tribes, and if so, to what extent do they act in concert, and under what leadership? Are personal conflicts common between men of the same tribe, and if so, what is their usual mode of proceeding?

2. *Weapons.*—What are their war weapons? are the same weapons used in war and the chase? What is the exact nature of their defensive armour, especially that described as being made of pieces of wood fastened together? Is the throwing-stick used in war? what is the accuracy, range, and penetration of a lance projected by this means? is there any evidence of its being a more ancient weapon than the bow? is it an indigenous weapon or derived from without? What are the difficulties in the construction of the bow from the absence of suitable elastic wood? is the practice of giving elasticity to the bow by means of sinews attached to it an independent invention or derived from the Asiatic Continent? what is the accuracy, range, and penetration of the bow?[1] In what manner are the performances of their weapons handed down from father to son, as is said to be the case? What is the exact meaning of the marks scored on their arrows and their weapons (with drawings of them)? Have they any means of giving a rotation to their arrows or other missile-weapons? Have they any regular system of training to the use of the bow and other weapons? At what age do the children commence the use of the bow? Are the Eskimos

---

[1] It appears desirable that some test of accuracy should be established. If the natives can be induced to shoot at a target, the distance of each shot from the point aimed at should be measured, added, and divided by the number of shots. The figure of merit obtained by this means would enable a comparison to be made with the shooting of other races. A target composed of grass bands, not less than six feet in diameter, might be used. Misses should be scored with a deviation of four feet; distances, fifty, one hundred, one hundred and fifty, and two hundred paces of thirty inches.

expert in throwing stones with the hand; and if so, how far can they throw with accuracy and force, and for what purpose do they throw stones? Is the bow drawn to the shoulder or the chest? is it held horizontally or vertically? Are the women trained to the use of weapons? What are the varieties of the weapons employed in different tribes and what is the cause of variation? to what extent do the weapons vary in form in each tribe? Have they anything resembling a standard, or state halbard, or fetish for war purposes, as suggested by Capt. Beechey? (Careful drawings and collections of all the varieties of weapons are very necessary.) To what extent have the natives abandoned their ancient arms, and taken to those of civilised nations introduced among them? Do they readily adopt European weapons?

3. *Leaders and Discipline.*—How are their leaders appointed? are they identical with the chiefs and Angekos? have they any marks or distinctions of dress (with drawings)? are they the strongest and most courageous? have they any rewards for warlike achievements? have they any subordinate leaders, and how are they appointed? have the chiefs any aids or runners to carry messages? What kind of discipline is preserved? Have they any punishments for offences in war? what is the function of the women in war? are any of the adult males reserved from war for employment in other duties that are necessary for the tribe, and if so, how is that arranged?

4. *Fortifications and Outposts.*—Have they any intrenchments, earth, or snow works or defensive pits, as described by Capt. Beechey, and if so, give plans and sections of them drawn to scale? Do they employ pitfalls in war or the chase, and if so, give plans and sections? Have they any knowledge of forming inundations for defensive purposes? Have they any use of stakes for defence, or stockades, or abatis? Do they employ caltraps (small spikes of wood fixed into the ground to wound the feet)? Do they ever build on raised piles for defence, as is practised in some parts of the N.W. Coast? Do they occupy isolated positions, or hills, or promontories for the defence of their villages? Do they fortify their villages or have they other strong places to resort to in case of attack which are not usually inhabited? Have they scouts and outposts, and are they arranged on any kind of regular system? Have they any special signals for war? do they employ special men on these duties?

5. *Supply.*—How do they supply themselves during war? does each man provide for himself or is there any general arrangement, and under what management? Are their proceedings much hampered by the difficulty of supply? How do they carry their food, water, and baggage?

6. *Causes and Effects of War.*—What are the chief causes of war?

Do feuds last long between tribes? How do they treat their prisoners? have they any special customs with regard to the first prisoner that falls into their hands? Do conquered tribes amalgamate? How are the women of the conquered tribes dealt with? How do they divide the spoil? Are their attacks always succeeded by retreat or do they follow up a victory? Is it likely that a knowledge of the arts, culture, &c., of other tribes has been spread by means of war? To what extent has the increase of the population been checked by wars? Has migration been promoted to any great extent by warlike expeditions?

ENQUIRIES RELATING to CERTAIN ARROW-MARKS and other SIGNS in USE AMONGST the ESKIMOS. By Col. A. LANE FOX.

1. Capt. Hall speaks of mysterious signs consisting of "particoloured patches sewn on to seal-skins, and hung up near the dwelling of the Angekok for the information of *strange* Innuit travellers, and to direct them what to do." Are these signs for *strange Innuit travellers* generally understood by the Eskimo race? what is their object and significance? are they generally understood by the people or only by the Angekos? Drawings and explanations of these signs would be desirable.

2. Sir Edward Belcher, in the 'Transactions of the Ethnological Society,' vol. i. p. 135, new series, gives his opinion that the Eskimos "are not without the means of recording events," and that "the use of notched sticks and working of the fingers has a deeper signification than mere numerals." What is the exact meaning of these marks? are they confined to particular tribes or common to the whole race? Drawings and collections of these notches would be desirable.

3. In our Ethnographical Museums identical marks upon horn-pointed arrows appear to be derived from different localities and at different times, so as to preclude the possibility of their having belonged to the same owner. Some of these marks appear to be pictographic, although consisting of straight lines representing a man or an animal; others are evidently not pictographic, and consist of a longitudinal line with other short lines branching from it, or an edge of the horn-point serves the purpose of the longitudinal lines, and the short lines are marked upon it. Their resemblance to Runes has been noticed. What is the exact meaning of each of these marks? are they the marks of the owner or do they record the performances of the weapon, or have they any other significance? are there similar marks upon other weapons and utensils or upon rocks? are they understood beyond the tribe? is there any probability of their having been derived from the Scandinavian settlers

in Greenland? Drawings and collections of these, and any other similar marks, with the exact meaning of each mark, would be desirable.

ENQUIRIES RELATING *to* DRAWING, CARVING, *and* ORNAMENTATION. By Col. A. LANE FOX.

Have the natives a natural aptitude for drawing? do they draw living animals in preference to other forms? are the heads of men and animals usually represented larger in proportion than the other parts of the body? Have they the least knowledge of perspective? Are the most distant objects drawn smaller than those nearer? are the more important personages or objects drawn larger than the others? Do their drawings represent imaginary animals or animals now extinct? Do they show any tendency to represent irregular objects, such as branching trees symmetrically so as to produce a conventional pattern? Are the drawings generally historical, or merely drawn for amusement or for ornament? Are events of different periods depicted in the same drawing? Have they any conventional modes of representing certain objects? Do they draw from nature or copy each other's drawings? Do they in copying from one another vary the forms through negligence, inability, or to save trouble, so as to lose sight of the original object and produce conventional forms, the nature of which is otherwise inexplicable? if so, it would be of great interest to obtain several series of such drawings, showing the gradual departure from the originals. Do they readily understand and appreciate European drawings? do they show any aptitude in copying European drawings? Do they draw with coloured earths besides the drawings engraved on bone? With what tools are these engravings made? Have they special artists who draw for the whole tribe or does each man ornament his own property? Do any of the natives show special talent for drawing, if so, in what direction does such talent show itself? Is drawing more practised in some tribes than others, and if so, does this arise from inclination or from traditional custom? Do they draw plans or maps? Do they understand European maps? At what age do the children commence drawing? are they encouraged to draw at an early age (a series of drawings of natives of different ages, from five or six upwards, would be interesting as a means of comparison with the development of artistic skill in Europeans)? Do they ornament with geometrical patterns, such as zigzags, concentric circles, contiguous circles, coils, spirals, punch-marks, lozenge patterns, herring-bone patterns, &c.? Do they use the continuous looped-coil pattern in ornamentation? Are such geometrical patterns in any case copies of mechanical contrivances, such as the binding of an arrow head, the strings supporting a vessel, &c., represented by incised lines? Are there any ancient drawings upon rocks, &c.? and, if so, in what respects do they differ from those of the existing natives? Copies to scale of any drawings which cannot be brought away would be very desirable.

5.—FURTHER ENQUIRIES *and* OBSERVATIONS *on* ETHNOLOGICAL QUESTIONS *connected with* ARCTIC EXPLORATION. By A. W. FRANKS, F.S.A., Keeper of Ethnography, &c., British Museum.

On reading over the enquiries suggested by the distinguished members of the Anthropological Institute, Dr. Barnard Davis, Mr. Tylor, Mr. Boyd Dawkins, and Col. Lane Fox, the following additional points of enquiry have suggested themselves:—

Anthropological Details.—Some uniform mode of measurement should be adopted, and careful instructions would no doubt be furnished by Dr. Barnard Davis. It would also be desirable to ascertain the strength of the natives in lifting and throwing weights, and pulling against weights, as compared with Europeans; also their speed in running.

Mental Qualities.—Evidences of quick understanding or the reverse. Habits of providence or the reverse. Knowledge of numeration and weights. Capability of understanding European pictures of animals, and especially of landscapes. Comprehension of the advantages of writing. Any knowledge of astronomy?

Marriage and Funeral Customs.—Is any ceremony observed with either sex on attaining puberty? Are wives obtained by courtship, capture, or purchase? if by the former, are there any surviving symbols of either of the two latter modes, as in Russia? At what age are marriages usually entered into? and are there any prohibited degrees of relationship? Are there any ceremonies at marriage or on childbirth? Is the name of the child ancestral? has it any special meaning? and is it changed at any time? How are the dead buried? are their weapons and food deposited with them? and if so, are they broken or rendered useless before being deposited? Is there any ceremony on receiving friends or strangers?

Arts and Manufactures.—Any particulars on these points will be of special value, as possibly illustrating prehistoric periods. How is the carving in ivory or bone executed? Is any method employed to soften the material? Have the ornamental designs on the implements any particular meaning? How are the skins tanned? are there any varieties in the fashions of dresses? and are these tribal or dependent on individual fancy? How is the sinew-thread made? Are labrets in use? and is tatooing employed by either sex? Is there any native explanation of either custom? It would be desirable to obtain the native names of the various tools, and to be especially attentive to the use of stone implements. Is meteoric iron employed for implements? and where is it obtained? The native names of metals employed? Are there special persons who manufacture a distinct class of objects or does each family supply its own wants? Is tobacco in use? where is it obtained? and is any other substance used with it or substituted for it? How are the tobacco-pipes made? and especially how are the bowls and stems bored?

Hunting and Fishing, &c.—The use of lures and stratagems. Are any tallies employed to record the number of animals killed? Is there any distinction in the form of paddles used by different sexes? do the rowers keep time?

Food.—Are any ceremonies used at their meals or feasts? Is there any offering to the deceased or to spirits? Is there any particular order in the succession of various kinds of food at such meals? Mode of feeding? especially as to the cutting off at the mouth the food. Do the teeth become much worn down by the nature of the food or the mode of eating?

Collections.—It is most desirable to make as complete a collection as possible of everything illustrating the Arctic tribes; for the intercourse with Europeans must in time modify or extinguish many of their peculiar implements, weapons, or dress, and it is believed that the Arctic races would furnish valuable illustrations of the condition of the ancient inhabitants of the South of France, &c., during the cave period. It would be well also to search in the walls and floors of ruined houses for stone and bone implements left by the former inhabitants. The specimens should be, as soon as possible, carefully labelled and marked; where marked by adhesive labels or by cards tied on, something should be written on the specimen itself, in ink or pencil, so that if the label should drop off or become detached there may be no doubt as to the specimen to which it belonged.

There is, however, a point of great importance which relates to the disposal of the collections when they are brought back. It has been too much the habit to consider such objects the property of the officers of the expedition, to be disposed of according to their wish. Should, however, such collections be made by a scientific expedition, there should be clear directions that it should be placed at the disposal of the Government to be deposited in the national museum, and the commander of the expedition should see that the main collection contains the best illustrations of the subject.

To show the evil effects of the contrary practice, it may be noticed that the greatest of English explorers, Captain Cook, must have made very large collections, as specimens obtained by him are to be found in many museums and private collections both in England and abroad. Unfortunately, the value of his specimens is much diminished by the absence of any proper account of the places from which they were derived ; and it is somewhat curious that although the British Museum is supposed to have the principal part of his collections, many of the finest specimens are not to be found there, but in other collections.

An instance connected with Arctic exploration may be noticed. In the well-known expedition in the *Blossom*, under Capt. Beechey, 1825–28, a number of specimens was obtained. Some of the specimens were given by Capt. Beechey to the Ashmolean Museum ; others were presented by the officers to Mr. Barrow, and are now in the British Museum. Sir Edward Belcher gave some of his specimens to the United Service Institution, which on the sale of a part of that museum were dispersed ; unfortunately they were not properly labelled, and their value is much impaired. The bulk of Sir Edward Belcher's collection has since been sold, and though by a fortunate accident some of the most interesting specimens have been secured for the Christy Collection, the value of the series as a whole is taken away. Others seem to have been given by Surgeon Collie to the Haslar Hospital, and on the breaking up of a portion of that museum were sent to the British Museum ; scarcely any of them were labelled, and it is only by accident that the probable origin of them has been traced. If a careful selection had been made at the time for the national collection, the manners, customs, and arts of the Western Eskimos would have received a full illustration.

6.—QUESTIONS *relating to the* PHYSICAL CHARACTERISTICS *of the* ESKIMOS, &c. By JOHN BEDDOE, M.D.

A. The following measurements should be obtained from as many adults of the two sexes as possible.

1. Stature: the best gotten by means of a graduated rod, in erect posture. Mention whether shoes are worn, and of what thickness.

2. Greatest length of head, from the eminence between the eye-brows ; with index or other callipers.

3. Greatest breadth of head, wherever found, with callipers.

4. Greatest breadth of zygomata, also with callipers.

5. Span—*i. e.*, distance between tips of middle fingers, arms being expanded.

6. Circumference of chest at nipple (in men).
7. Ditto after full expansion by forced inspiration (in men).
8. Circumference of thigh at fork.
9. Distance from fork to ground.
1, 6, and 9, are most important.

B. The colours of *hair*, *eyes*, and *skin*, may be best expressed by means of Broca's scale; but in its absence the

1. *Eyes* may be designated as light (blue, light grey, light green), neutral (dark grey, dark green, yellowish grey), or dark hazel, brown).

2. *Hair* as red, fair, brown, dark brown, rusty black, or coal-black.

3. It should be noted whether there is any beard, and, if so, of what colour, or whether it is extirpated.

4. Is grey hair observed?

5. Or baldness?

6. or the *arcus senilis?*

7. Is the hair lighter in children than in adults?

8. Is the body less hairy than in Europeans?

C. 1. What is the temperature of the body, taken with a " clinical thermometer" kept in the axilla fully five minutes? This should be observed in four or five persons.

2. Does the hand appear to be notably smaller than in Europeans?

For use in the observations above, a graduated rod, six feet long, with a sliding cross-piece, index callipers, graduated tapes, and a clinical thermometer will be desirable.

7.—FURTHER ETHNOLOGICAL ENQUIRIES, *more especially connected with the* WESTERN ESKIMOS. By WILLIAM TURNER, Professor of Anatomy, University of Edinburgh.

1. Should the expedition visit the western part of the north coast of America, it would be very desirable to ascertain if any traditions linger amongst the Eskimo tribes of a migration of their ancestors across Behring Straits.

2. It would also be desirable to ascertain if any communication takes place between the Eskimos and the most northerly tribes of North American Indians, either for purposes of trade or war; or if the Eskimos or Indian tribes intermarry.

3. Collections of crania of the tribes occupying the land on the

x

eastern and western sides of Behring's Straits would be of great
value. Careful notes should also be taken of the physical characteristics of the people, of their habits and modes of life, their tools,
weapons, &c.

4. A collection of crania from the district around Kotzebue Sound
would be also prized, as there is reason to think, from a few specimens already in this country, that the cranial configuration of the
people of this region differs from that of the tribes on the eastern
side of the American continent.

8.—INSTRUCTIONS SUGGESTED BY CAPT. BEDFORD PIM, R.N.

1. Make full inquiries as to the shape, length, breadth, depth,
and capacity of the baidars; the covering, the lashing, size of the
ribs and timbers, and the dimensions of the paddles.

2. How many persons can the baidar carry? with how much
weight inside will they float when swamped?

3. What amount of provisions for its occupants can the baidar
carry? what is the nature of those provisions, and how many days
will they last?

4. What is the utmost speed of a baidar under paddles, paddles
and sail (if any), or sail (if any) alone?

5. How many miles can be paddled in four hours? ditto eight
hours? ditto twelve hours, with the view to arrive at the length of
a day's journey?

6. These questions to apply equally to the kayak.

7. Especially make inquiries with reference to the capability of
the baidar, or of two kayaks lashed together, to cross from Labrador
to Greenland; and their ability to encounter heavy weather.

8. Also if women can paddle the kayak as well as the men.

9. Make particular inquiries about the weapons of the chase used
both on land and water.

4 1 8	8

LONDON: PRINTED BY WILLIAM CLOWES AND SONS, STAMFORD STREET,
AND CHARING CROSS.

www.ingramcontent.com/pod-product-compliance
Lightning Source LLC
Chambersburg PA
CBHW021506210326
41599CB00012B/1155